CW00550856

BASIC WING
and
AIRFOIL THEORY

The trailing wing-tip vortices of a B-25 Mitchell pick up a smoke screen being laid in the Southwest Pacific. (*Courtesy North American Aircraft.*)

BASIC WING
and
AIRFOIL THEORY

ALAN POPE

DOVER PUBLICATIONS, INC.
MINEOLA, NEW YORK

Bibliographical Note

This Dover edition, first published in 2009, is an unabridged republication of the work originally published in 1951 by McGraw-Hill Book Company, Inc., New York.

Library of Congress Cataloging-in-Publication Data

Pope, Alan, 1913–
    Basic wing and airfoil theory / Alan Pope.
        p. cm.
    Originally published: New York : McGraw-Hill, 1951.
    Includes bibliographical references and index.
    ISBN-13: 978-0-486-47188-4
    ISBN-10: 0-486-47188-8
    1. Aerofoils. 2. Airplanes—Wings. 3. Aerodynamics. I. Title.

TL574.A4P58 2009
629.134'32—dc22

2008044627

Manufactured in the United States of America
Dover Publications, Inc., 31 East 2nd Street, Mineola, N.Y. 11501

PREFACE

Books in the field of wing and airfoil theory have been written either at a very elementary level for the college junior or at a high level for the advanced student. Few, if any, papers have been presented at a level directly suitable for the college senior or first-year graduate student. Noting this situation in 1941, the author started a course at the "between" level, which has finally emerged as this textbook after some delay because of wartime exigencies. The material has been used in mimeograph form at the Georgia Institute of Technology in a one-year course.

It was impossible to avoid limited use of some mathematical tools not customarily found at undergraduate levels, and hence two chapters (the first two) have been provided to afford the student a very light covering of vector analysis and complex variables. The more advanced students may start with Chapter 3.

While this book is an integrated whole and represents what the author feels is needed in a low-level course on wing and airfoil theory, the general aerodynamicist may prefer to omit Chapter 7 (Thin-airfoil Theory) and Chapter 8 (Arbitrary Airfoil Theory) in favor of more attention to the flow field for the entire airplane (Chapters 9 through 13). The omission of these two chapters will in no way interfere with a clear understanding of the remainder of the book.

For the instructor's information, this book has been used in two forms: (1) Followed straight through, three hours a week for a school year, with all equations derived by the students, all problems worked, and a maximum of use made of the references. (2) As a considerably shorter course in which Chapter 1 (Vector Analysis) was omitted, along with all references to vectors in the text and parallel reading. In this case more attention is given to the use of the relations, and less to their derivation. It is hoped that the book will therefore be useful at two levels.

So much of the spadework in this textbook was completed by all the graduate students with whom I have had the pleasure of working that I hesitate to single out any for special thanks. Considering the risk, however, I feel it necessary to mention four: A. L. Ducoffe, Melvin Towsley, R. E. Wilson, and J. R. Cumberpatch. To them, and to the other graduate and undergraduate students who have labored with this manuscript in mimeograph form, go my heartfelt thanks.

In addition I wish to thank Dr. M. J. Thompson of the University of
Texas for permission to draw upon his lecture notes in part of Chapter 7.
Any errors in this book are, of course, my responsibility.

ALAN POPE

ATLANTA, GA.
  February, 1951

CONTENTS

CHAPTER 1. VECTOR ANALYSIS

CHAPTER 2. COMPLEX VARIABLES AND FOURIER SERIES

CHAPTER 3. THE STREAM FUNCTION

CHAPTER 7. THIN-AIRFOIL THEORY

CHAPTER 8. THICK-AIRFOIL THEORIES

CHAPTER 9. THE FINITE WING

CHAPTER 10. THE MONOPLANE WING

CHAPTER 11. THE FIELD ABOUT THE WING

ABBREVIATIONS

In order to shorten the context, the following standard abbreviations are employed:

JAS   Journal of the Aeronautical Sciences

JRAS   Journal of the Royal Aeronautical Society (British)

NACA   National Advisory Committee for Aeronautics

PRS   Proceedings of the Royal Society of London (British)

R & M   Reports and Memoranda (of the British Air Research Committee)

RTP   Research of Technical Progress (British)

TM   Technical Memorandum of the NACA

TN   Technical Note of the NACA

TR   Technical Report of the NACA

WRL   Wartime Report (Langley) of the NACA

LIST OF SYMBOLS

a in potential theory, the semi-major axis of a Rankine oval; in transformation theory, the radius of the circle to be transformed; in wing theory, the slope of the lift curve; in the Biot-Savart derivation, the distance from a point to an elemental area of the vortex sheet

a_1 a vector of unit length

a_0 in wing theory, the slope of the lift curve for infinite aspect ratio, usually taken as 2π per radian

A_n a general term of a series

A_1 the first term of the A_n series; in thin-airfoil theory $= -2\alpha_{ZL}$; in wing theory $= \dfrac{C_L}{\pi \text{ A.R.}}$

A.R.   aspect ratio $= \dfrac{b^2}{S} = \dfrac{4s^2}{S}$

$A_x\ A_y\ A_z$ the components of vector $\mathbf{A}$ along the x, y, and z axes

b in potential theory, the semi-minor axis of a Rankine oval; in wing theory, the wing span, ft

b_1 the distance between the lateral tunnel walls, ft

C wind-tunnel-test section area, sq ft

c in wing theory, the wing chord, ft

c_1 in airfoil theory, the distance from the origin to the rear stagnation point of the circle to be transformed; approximately one-fourth of the chord of the ensuing airfoil

$c_C,\ C_C$ chord force coefficient

c_{do} section profile drag coefficient

C_{DO} wing profile drag coefficient

$c_m,\ C_m$ moment coefficients

$c_N,\ C_N$ normal force coefficients

c_l section lift coefficient

C_L wing lift coefficient

d airfoil maximum thickness, ft

e the natural logarithm base

E the edge correction factor

f the camber factor $= 2m$

h in the Biot-Savart derivation, the distance from a vortex to point P; in tunnel-wall studies the height of the test section, ft

i complex notation for $\sqrt{-1}$

i, j, k unit vectors along the x, y, z axes

I the imaginary part of an expression

k a parameter in the Kármán-Trefftz theory $= \dfrac{a}{c_1}$

l in vectors, the cosine of the angle between a vector and the x axis; in wing theory, the distance behind the lifting line, ft

m in vectors, the cosine of the angle between a vector and the y axis; in transformation theory, the distance from the origin to the center of the circle to be transformed

M a moment; its subscript indicates the axis about which it was taken

n in vectors, the cosine of the angle between a vector and the z axis; in series, any whole number; in the Kármán-Trefftz theory, a number slightly less than 2.0

p in series, any whole number (usually used only when n and m have been previously assigned); a pressure, lb per sq ft; in the Theodorsen theory, a function of x, y, and a

q in potential theory, a velocity ft per sec; in wing theory, the dynamic pressure, lb per sq ft

q_r outward radial velocity, ft per sec

q_θ counterclockwise tangential velocity, ft per sec

q_z a velocity in the z plane, ft per sec

q_ζ a velocity in the ζ plane, ft per sec

Q flow quantity from a source, sq ft per sec

R ae^{ψ_0}

Re   the real part of an expression

s one-half the wing span, ft; in potential theory, the distance from a source or sink to the y axis, ft

s' one-half the vortex span far aft of a wing, ft

S the wing area, sq ft

u a horizontal velocity, ft per sec

v in wing theory, a velocity in the starboard direction, ft per sec

v_n a vertical velocity, ft per sec

V the free-stream velocity

w in potential theory, the potential function, sq ft per sec; in wing theory, a downward velocity, ft per sec

w_0 the downwash at the lifting line for an elliptic wing

W the weight of an airplane, lb

x the horizontal distance from the y axis

y in airfoil theory, the distance above the x axis; in wing theory, the distance from the plane of symmetry to starboard

$\bar{y}$ nondimensional distance from a point to the plane of symmetry $= \dfrac{y}{b/2}$

z in wing theory, the distance below the longitudinal axis; in complex notation $= x + iy$

α (alpha)   an angle; in vectors, the angle between a vector and the x axis; in aerodynamics, the angle between an arbitrary chord line and the relative wind; sometimes the angle between the zero-lift chord and the relative wind

α_i the induced angle at a wing

α_{ZL} the angle of zero lift

β (beta)   an angle; in vectors, the angle between a vector and the y axis; in airfoil theory, the angle between the x axis and the line from the rear stagnation point to the center of the circle to be transformed

γ (gamma)   an angle; in vectors, the angle between a vector and the z axis; in Multhopp's theory (Chap. 12), the nondimensional circulation $= \dfrac{\Gamma}{bV}$

Γ (gamma)   the circulation, sq ft per sec

δ (delta)   a small positive factor that increases the induced drag to allow for nonelliptic loading; in airfoil theory, the angle between the vector from the origin to the center of the z circle and the x axis

ϵ (epsilon)   in Joukowski airfoil theory, a thickness parameter $= \dfrac{m}{c_1}$; in wing theory, the downwash angle $\dfrac{w}{V}$; in Theodorsen's theory $= \varphi - \theta$

ζ (zeta)   in Theodorsen's theory $= x + iy$; in Joukowski airfoil theory $= \xi + i\eta$

η (eta)   vertical distance from the axis; also efficiency; in Chap. 12, the distance from a vortex to the plane of symmetry, ft

θ (theta)   the angle between the radius vector to a point and the x axis

θ' the angle between the relative wind and the line joining the center of the z circle and a point

θ_1 the angle between a radius of the z circle and the x axis

λ (lambda)   the "local aspect ratio" $= \dfrac{b}{c}$

μ (mu)   in potential theory, the doublet strength $= 2Qs$; in airfoil theory, the center of the z circle; in wing theory, a wing parameter $= \dfrac{a_0 c}{8s}$

ξ (xi)   horizontal distance from the η axis

ρ (rho)   the density, slug per cu ft; in the Theodorsen theory, the nose radius of the airfoil

τ (tau)   a small positive factor that corrects the induced angle for nonelliptic loading

φ (phi)   in the Theodorsen theory, the argument of a point on the z circle

ϕ (phi)   the velocity potential, sq ft per sec; in the Kármán-Trefftz theory, the apex angle in the untransformed plane

ψ (psi)   in the Theodorsen theory, the exponent of the e function used to vary the radius of the z circle; in potential theory, the stream function, sq ft per sec

ω (omega)   an angular velocity, radians per sec

CHAPTER 1

VECTOR ANALYSIS

1.1. Vectors. On numerous occasions in aerodynamic theory, the use of vectors and vector notation greatly reduces the mathematical steps required by ordinary means. Actually this text does not lean heavily upon vector notation, but in order that the student may read additional works in this field it is quite necessary to have a good grasp of vector fundamentals.

A quantity may be *vector* or *scalar*. Scalar quantities have magnitude only, as, for example, *temperature, density,* or *pressure.* We shall indicate scalars by simple, unadorned letters such as A. Vector quantities have direction as well as magnitude: examples are *forces, velocities, acceleration,* and *weight.*

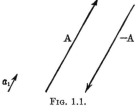

Fig. 1.1.

Vectors are indicated by boldface type or by a bar $\bar{A}$ or a wiggle A.

A vector may be considered as a whole **A** or as a scalar quantity A with a direction indicated by a unit vector a_1, as

$$\mathbf{A} = a_1 A \tag{1.1}$$

and although a_1 is a vector, it has no identifying mark and must be recognized by its place in an equation.

1.2. Components of a Vector. The components of a vector **A** making angles α, β, and γ with the x, y, and z axes are, respectively, $A \cos \alpha$, $A \cos \beta$, and $A \cos \gamma$. Letting directions along the x, y, and z axes be indicated by *unit vectors i, j,* and k, we have

$$\mathbf{A} = iA \cos \alpha + jA \cos \beta + kA \cos \gamma$$

or, introducing another type of nomenclature,

$$\mathbf{A} = iA_x + jA_y + kA_z \tag{1.2}$$

From Fig. 1.2 we see that

$$|\mathbf{A}| = |\sqrt{A_x{}^2 + A_y{}^2 + A_z{}^2}| \tag{1.3}$$

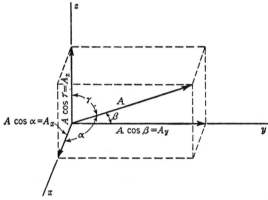

FIG. 1.2

1.3. Addition of Vectors. From Fig. 1.3 we see that

$$\mathbf{A} + \mathbf{B} = i(A_x + B_x) + j(A_y + B_y) + k(A_z + B_z) \qquad (1.4)$$

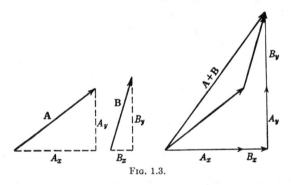

FIG. 1.3.

Example 1.1. Find the resultant and the length of the resultant of

$$\mathbf{P}_1 + \mathbf{P}_2 + \mathbf{P}_3$$

if

$$\mathbf{P}_1 = 3i + j - k$$
$$\mathbf{P}_2 = 4i + 2j + 3k$$
$$\mathbf{P}_3 = 3i - 2j + 4k$$
$$\mathbf{R} = (3 + 4 + 3)i + (1 + 2 - 2)j + (-1 + 3 + 4)k = 10i + j + 6k$$
$$|\mathbf{R}| = \sqrt{137}$$

1.4. Subtraction of Vectors. It also follows from Fig. 1.3 that

$$\mathbf{A} - \mathbf{B} = i(A_x - B_x) + j(A_y - B_y) + k(A_z - B_z) \qquad (1.5)$$

or

$$\mathbf{A} - \mathbf{B} = \mathbf{A} + (-\mathbf{B}) \qquad (1.6)$$

1.5. Vector Cross Product. The *cross product* of two vectors is defined by

$$\mathbf{A} \times \mathbf{B} = \epsilon AB \sin \theta \tag{1.7}$$

where ϵ is a unit vector whose direction is perpendicular to the plane of $\mathbf{A}$ and $\mathbf{B}$ in the right-hand screw direction of $\mathbf{A}$ rotated toward $\mathbf{B}$ and θ is the angle between $\mathbf{A}$ and $\mathbf{B}$.

Since $\sin 0° = 0$,

$$i \times i = j \times j = k \times k = 0 \tag{1.8}$$

Also, since $\sin 90° = 1.0$,

$$i \times j = k; \qquad j \times k = i; \qquad k \times i = j \tag{1.9}$$

Let

$$\mathbf{P} = iP_x + jP_y + kP_z$$

and

$$\mathbf{Q} = iQ_x + jQ_y + kQ_z$$

Then the cross product $\mathbf{P} \times \mathbf{Q}$ from Eqs. (1.2), (1.7), and (1.9) is

$$\begin{aligned}
\mathbf{P} \times \mathbf{Q} &= (iP_x + jP_y + kP_z) \times (iQ_x + jQ_y + kQ_z) \\
&= i(P_yQ_z - P_zQ_y) + j(P_zQ_x - P_xQ_z) + k(P_xQ_y - P_yQ_x) \quad (1.10)
\end{aligned}$$

Consider the determinant

$$D = \begin{vmatrix} i & j & k \\ P_x & P_y & P_z \\ Q_x & Q_y & Q_z \end{vmatrix} \tag{1.11}$$

Expanding D in the usual manner, we get

$$D = i(P_yQ_z - P_zQ_y) + j(P_zQ_x - P_xQ_z) + k(P_xQ_y - P_yQ_z) \tag{1.12}$$

Since Eqs. (1.10) and (1.12) are identical, we conclude that the cross product may be directly represented by a determinant as

$$\mathbf{P} \times \mathbf{Q} = \begin{vmatrix} i & j & k \\ P_x & P_y & P_z \\ Q_x & Q_y & Q_z \end{vmatrix} \tag{1.13}$$

It is well to emphasize that the cross product of two vectors is a vector itself. (Later we shall introduce a vector product that is a scalar.)

It follows from the above that

$$\mathbf{A} \times \mathbf{B} = -\mathbf{B} \times \mathbf{A} \qquad (1.14)$$

Example 1.2.   Find $\mathbf{A} \times \mathbf{B}$ if $\mathbf{A} = 3i + 2j - 4k$ and $\mathbf{B} = 6i - 2j + 8k$.

$$\mathbf{A} \times \mathbf{B} = \begin{vmatrix} i & j & k \\ 3 & 2 & -4 \\ 6 & -2 & +8 \end{vmatrix}$$
$$= i(16 - 8) + j(-24 - 24) + k(-6 - 12)$$
$$= 8i - 48j - 18k$$

1.6. Cross-product Significance
1. If

$$\mathbf{A} \times \mathbf{B} = 0 \qquad (1.15)$$

$\mathbf{A}$ and $\mathbf{B}$ are parallel.

2.

$$(\mathbf{P} + \mathbf{Q}) \times \mathbf{R} = \mathbf{P} \times \mathbf{R} + \mathbf{Q} \times \mathbf{R} \qquad (1.16)$$

The validity of Eq. (1.16) may be checked by writing $\mathbf{P} + \mathbf{Q}$ as $i(P_x + Q_x)$, etc., and crossing it into $R = iR_x + jR_y + kR_z$ and comparing the result obtained with the expansions of $\mathbf{P} \times \mathbf{R} + \mathbf{Q} \times \mathbf{R}$.

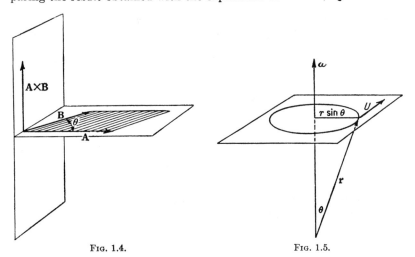

Fig. 1.4.   Fig. 1.5.

3. *Cross product as an area.*   By definition,

$$\mathbf{A} \times \mathbf{B} = \epsilon AB \sin \theta$$

and

$$B \sin \theta = \text{altitude of shaded parallelogram} \qquad (\text{Fig. 1.4})$$

$$\therefore |\mathbf{A} \times \mathbf{B}| = \text{area of shaded parallelogram}$$

4. *Cross product for rotation.* Let a particle rotate (see Fig. 1.5) with an angular velocity ω, in a circle of radius $r \sin \theta$. Its linear velocity

$$U = \epsilon \omega r \sin \theta = \omega \times r \qquad (1.17)$$

5. *Cross products for couples.* The moment about O is (Fig. 1.6)

$$M = R_2 \sin \theta_2 \cdot F - R_1 \sin \theta_1 \cdot F$$
$$M = (R_2 - R_1) \times F$$

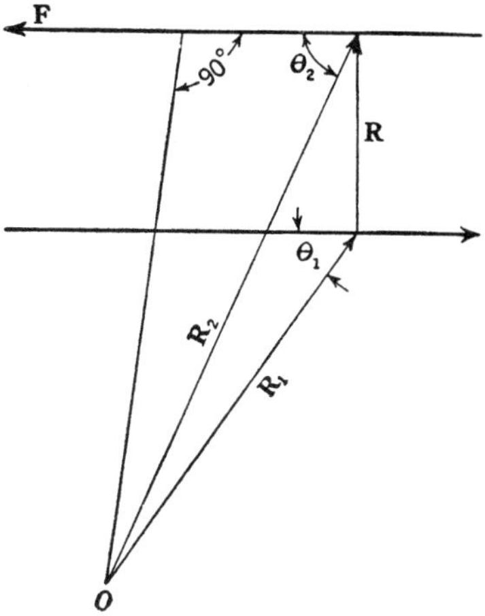

But from Fig. 1.6

$$R_2 - R_1 = R$$

Hence

$$M = R \times F \qquad (1.18)$$

It may be noted that the disappearance of the position vectors R_1 and R_2 substantiates the well-known principle that the moment of a couple is the same about any point.

1.7. Dot, or Scalar, Product. The dot, or scalar, product of two vectors is defined by the relation

$$A \cdot B = AB \cos \theta \qquad (1.19)$$

where θ = angle between A and B.

We see immediately that if $A \cdot B = 0$, A and B are perpendicular.

$B \cos \theta$ is called the *component* of B along A, and the dot product can hence represent the work done by B along A.

Since the dot product results in a scalar

$$A \cdot B = B \cdot A \qquad (1.20)$$

and

$$i \cdot i = j \cdot j = k \cdot k = 1 \qquad (1.21)$$
$$i \cdot j = j \cdot k = k \cdot i = 0 \qquad (1.22)$$

Example 1.3. Show that

$$A \cdot B = A_x B_x + A_y B_y + A_z B_z$$

Writing

$$A = iA_x + jA_y + kA_z$$

and

$$B = iB_x + jB_y + kB_z$$

and applying Eqs. (1.21) and (1.22) as we take the dot product, we have

$$A \cdot B = A_x B_x + A_y B_y + A_z B_z \qquad (1.23)$$

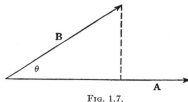

FIG. 1.6.

FIG. 1.7.

1.8. Division of Vectors.   From Fig. 1.8

$$\mathbf{A} \cdot \mathbf{B} = \mathbf{A} \cdot \mathbf{C}$$

If we could directly divide a vector equality, we could cancel out **A**, leaving

$$\mathbf{B} = \mathbf{C}$$

an obviously erroneous conclusion.   We shall therefore not include vector division.

1.9. Triple Product, $\mathbf{A} \cdot \mathbf{B} \times \mathbf{C}$.   Recalling that the dot product $\mathbf{A} \cdot \mathbf{B}$ yields a scalar and the cross product $\mathbf{B} \times \mathbf{C}$ yields a vector, we conclude that to form a triple product $\mathbf{A} \cdot \mathbf{B} \times \mathbf{C}$ we must first take the cross

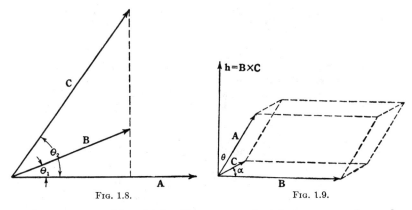

FIG. 1.8.   FIG. 1.9.

product as we could not take the cross product between a scalar and a vector.   That is,

$$\mathbf{A} \cdot \mathbf{B} \times \mathbf{C} = \mathbf{A} \cdot (\mathbf{B} \times \mathbf{C}) \tag{1.24}$$

Consider the parallelepiped in Fig. 1.9.   It is seen that $\mathbf{B} \times \mathbf{C} = \mathbf{h}$, the area of the base; and $\mathbf{h} \cdot \mathbf{A} = \mathbf{A} \cdot \mathbf{B} \times \mathbf{C}$ is the volume of the parallelepiped since $A \cos \theta$ is its altitude.

We further note that $\mathbf{A} \cdot \mathbf{B} \times \mathbf{C}$ is a scalar.   From Fig. 1.9

$$\mathbf{A} \cdot \mathbf{B} \times \mathbf{C} = \mathbf{C} \cdot \mathbf{A} \times \mathbf{B} = \mathbf{B} \cdot \mathbf{C} \times \mathbf{A} \tag{1.25}$$

Other than cyclic permutation changes the sign.   If **A**, **B**, and **C** lie in a plane, the volume of the parallelepiped is zero and $\mathbf{A} \cdot \mathbf{B} \times \mathbf{C} = 0$.

The actual evaluation of the triple product can be made through the use of a simple determinant as follows: From Eq. (1.13)

$$\mathbf{B} \times \mathbf{C} = \begin{vmatrix} i & j & k \\ B_x & B_y & B_z \\ C_x & C_y & C_z \end{vmatrix} = i(B_y C_z - B_z C_y) \cdots$$

Then

$$\mathbf{A} \cdot \mathbf{B} \times \mathbf{C} = (iA_x + jA_y + kA_z) \cdot [i(B_yC_z - B_zC_y) + \cdots]$$
$$= A_xB_yC_z - A_xB_zC_y \cdots$$

Expanding

$$\begin{vmatrix} A_x & A_y & A_z \\ B_x & B_y & B_z \\ C_x & C_y & C_z \end{vmatrix} = A_xB_yC_z - A_xB_zC_y \cdots$$

and hence

$$\mathbf{A} \cdot \mathbf{B} \times \mathbf{C} = \begin{vmatrix} A_x & A_y & A_z \\ B_x & B_y & B_z \\ C_x & C_y & C_z \end{vmatrix} \tag{1.26}$$

1.10. Triple Vector Product, $\mathbf{A} \times (\mathbf{B} \times \mathbf{C})$. The triple vector product

$$\mathbf{A} \times (\mathbf{B} \times \mathbf{C}) = \mathbf{B}(\mathbf{A} \cdot \mathbf{C}) - \mathbf{C}(\mathbf{A} \cdot \mathbf{B}) \tag{1.27}$$

where the operations in the parentheses are to be performed first. Proof of Eq. (1.27) is left to the student.

1.11. Direction Cosines. The direction cosines are the cosines of the angles between a given vector and the three axes. They are designated l, m, and n for the x, y, and z axes, respectively.

For the vector

$$\mathbf{R} = iR_x + jR_y + kR_z \tag{1.28}$$

$$l = \frac{R_x}{|\mathbf{R}|}; \qquad m = \frac{R_y}{|\mathbf{R}|}; \qquad n = \frac{R_z}{|\mathbf{R}|} \tag{1.29}$$

Example 1.4. Find the direction cosines of the vector $\mathbf{A} = 3i - 4j + 2k$.

$$|\mathbf{A}| = |\sqrt{3^2 + (-4)^2 + 2^2}| = 5.38$$

$$l = \frac{3}{5.38} = 0.558; \qquad \alpha = 56.2°$$

$$m = \frac{-4}{5.38} = -0.743; \qquad \beta = 138.0°$$

$$n = \frac{2}{5.38} = 0.372; \qquad \gamma = 68.2°$$

1.12. Angle between Vectors. The definition of the dot product leads directly to the angle between two vectors, as follows: From Eq. (1.19)

$$\mathbf{A} \cdot \mathbf{B} = AB \cos \theta \tag{1.30}$$

and hence

$$\cos \theta = \frac{A_xB_x + A_yB_y + A_zB_z}{|\mathbf{A}|\,|\mathbf{B}|} \tag{1.31}$$

Example 1.5. Find the angle between vector $\mathbf{A} = 2i + j - 3k$ and vector $\mathbf{B} = -3i - j + k$.

$$\mathbf{A} \cdot \mathbf{B} = A_x B_x + A_y B_y + A_z B_z = -6 - 1 - 3 = -10$$
$$|\mathbf{A}| = |\sqrt{2^2 + 1^2 + (-3)^2}| = |\sqrt{14}|$$
$$|\mathbf{B}| = |\sqrt{(-3)^2 + (-1)^2 + 1^2}| = |\sqrt{11}|$$
$$\cos \theta = \frac{-10}{\sqrt{154}} = -0.806, \qquad \theta = 143°42'$$

1.13. Vector Derivatives. The derivatives of vectors follow quite directly. Let
$$\mathbf{r} = ix + jy + kz$$
Then
$$d\mathbf{r} = i\,dx + j\,dy + k\,dz \tag{1.32}$$
and
$$\frac{d\mathbf{r}}{dt} = i\frac{dx}{dt} + j\frac{dy}{dt} + k\frac{dz}{dt}$$

1.14. Derivative of the Cross Product. The derivative of the cross product
$$\frac{d\,(\mathbf{A} \times \mathbf{B})}{dt} = \mathbf{A} \times \frac{d\mathbf{B}}{dt} + \frac{d\mathbf{A}}{dt} \times \mathbf{B} \tag{1.33}$$
is developed as follows: From Eq. (1.10)
$$\mathbf{A} \times \mathbf{B} = iA_y B_z - iA_z B_y \cdots$$
and hence
$$\frac{d(\mathbf{A} \times \mathbf{B})}{dt} = iA_y\frac{dB_z}{dt} + iB_z\frac{dA_y}{dt} - iA_z\frac{dB_y}{dt} - iB_y\frac{dA_z}{dt} \cdots \tag{1.34}$$
Consider
$$\mathbf{A} \times \frac{d\mathbf{B}}{dt} = iA_y\frac{dB_z}{dt} - iA_z\frac{dB_y}{dt} \cdots \tag{1.35}$$
and
$$\frac{d\mathbf{A}}{dt} \times \mathbf{B} = iB_z\frac{dA_y}{dt} - iB_y\frac{dA_z}{dt} \cdots \tag{1.36}$$
It will be seen that Eq. (1.35) plus Eq. (1.36) equals Eq. (1.34), thus verifying Eq. (1.33).

1.15. Derivative of the Dot Product. In a manner similar to that of Sect. 1.14, the following equation may be developed:
$$\frac{d(\mathbf{P} \cdot \mathbf{Q})}{dt} = \mathbf{P} \cdot \frac{d\mathbf{Q}}{dt} + \frac{d\mathbf{P}}{dt} \cdot \mathbf{Q} \tag{1.37}$$
Proof of Eq. (1.37) is left to the student.

1.16. Fields. If each point of a region has a value of some quantity, the region is called a *field*. If the quantity is scalar (for example, temperature), the field is scalar; if the quantity has direction (velocity, force, etc.), the field is a vector field.

If f is a scalar having a value $P(x, y, z)$ at each point, and $\mathbf{r}$ is the vector to that point from the origin, then f is a function of $\mathbf{r}$ in that for each value of $\mathbf{r} = ix + jy + kz$ there corresponds a point P and a particular value for f. It can be seen that the use of the position vector $\mathbf{r}$ enables scalar problems to be reduced to vector notation.

If f is a vector, then f is a vector function of $\mathbf{r}$ in the same manner.

1.17. The Gradient del. Suppose a function

$$\phi = \phi(x, y, z)$$

has a position vector

$$\mathbf{r} = ix + jy + kz$$

Then the total derivative of ϕ

$$d\phi = \frac{\partial \phi}{\partial x}\, dx + \frac{\partial \phi}{\partial y}\, dy + \frac{\partial \phi}{\partial z}\, dz$$

may be written as

$$d\phi = \left(i\, \frac{\partial \phi}{\partial x} + j\, \frac{\partial \phi}{\partial y} + k\, \frac{\partial \phi}{\partial z}\right) \cdot (i\, dx + j\, dy + k\, dz)$$

and the quantity in the first parentheses is called the gradient of

$$\phi \ (\equiv \text{del } \phi \equiv \nabla \phi)$$

Quantities which act upon others are called *operators*, and the operator del is hence

$$\nabla = i\, \frac{\partial}{\partial x} + j\, \frac{j}{\partial y} + k\, \frac{\partial}{\partial z} \quad (1.38)$$

By definition it operates only on quantities on its right. Equation (1.38) shows that $\nabla \phi$ is a vector that has in each direction a component equal to the derivative in that direction. It is shown in more advanced texts that $\nabla \phi$ itself points in the direction of the maximum

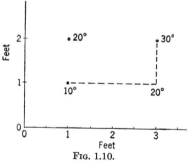

Fig. 1.10.

rate of change of ϕ, and its magnitude is that of the maximum rate of change.

For example, consider a field in which the temperature varies as in Fig. 1.10. The change along x is 5° per foot and along y is 10° per foot. Hence $\dfrac{\partial T}{\partial x} = 5$, and $\dfrac{\partial T}{\partial y} = 10$. The gradient of T, $\nabla T = 5i + 10j$.

1.18. Applications of the Gradient. By definition, ∇ is both an operator and a vector; and also by definition it operates only on quantities on its *right*. For example,

$$\nabla\phi = i\frac{\partial\phi}{\partial x} + j\frac{\partial\phi}{\partial y} + k\frac{\partial\phi}{\partial z} \tag{1.39}$$

Del ϕ is a vector, although ϕ is a scalar. We may take $\nabla \cdot \mathbf{V}$ or $\nabla \times \mathbf{V}$, but, for example, $\nabla \cdot \mathbf{V}$ does not equal $\mathbf{V} \cdot \nabla$ as ∇ operates only on terms on its right side.

Example 1.6. Show that $\nabla \times \mathbf{r} = 0$.

$$\nabla \times \mathbf{r} = \begin{vmatrix} i & j & k \\ \frac{\partial}{\partial x} & \frac{\partial}{\partial y} & \frac{\partial}{\partial z} \\ x & y & z \end{vmatrix} = i\left(\frac{\partial z}{\partial y} - \frac{\partial y}{\partial z}\right) + \cdots = i(0-0) = 0$$

which holds as long as x, y, z are independent variables.

Additional applications of del will be considered in Sect. 1.20.

1.19. Successive Applications of the Gradient; the Laplacian. At a later time it will be advantageous to use successive applications of the gradient such as $\nabla \cdot \nabla\phi$. This term may be expanded as follows:

$$\nabla \cdot \nabla\phi = \left(i\frac{\partial}{\partial x} + j\frac{\partial}{\partial y} + k\frac{\partial}{\partial z}\right) \cdot \left(i\frac{\partial\phi}{\partial x} + j\frac{\partial\phi}{\partial y} + k\frac{\partial\phi}{\partial z}\right)$$
$$= \frac{\partial^2\phi}{\partial x^2} + \frac{\partial^2\phi}{\partial y^2} + \frac{\partial^2\phi}{\partial z^2}$$

Defining

$$\nabla^2 = \nabla \cdot \nabla = \frac{\partial^2}{\partial x^2} + \frac{\partial^2}{\partial y^2} + \frac{\partial^2}{\partial z^2}$$

we call ∇^2 the Laplacian. Note that ∇ is a vector operator, while ∇^2 is a scalar operator.

1.20. Important Exercises. Students should demonstrate the validity of the following:

1. $\nabla \cdot (\mathbf{V}_1 + \mathbf{V}_2) = \nabla \cdot \mathbf{V}_1 + \nabla \cdot \mathbf{V}_2$
2. $\nabla \times (\mathbf{V}_1 + \mathbf{V}_2) = \nabla \times \mathbf{V}_1 + \nabla \times \mathbf{V}_2$
3. $\nabla \cdot \phi\mathbf{U} = \phi\nabla \cdot \mathbf{U} + \mathbf{U} \cdot \nabla\phi$
4. $\nabla \times \phi\mathbf{U} = \phi\nabla \times \mathbf{U} + \nabla\phi \times \mathbf{U}$
5. $\nabla \cdot \mathbf{U} \times \mathbf{V} = \mathbf{V} \cdot \nabla \times \mathbf{U} - \mathbf{U} \cdot \nabla \times \mathbf{V}$
6. $\nabla \times \mathbf{U} \times \mathbf{V} = \mathbf{V} \cdot \nabla\mathbf{U} - \mathbf{U} \cdot \nabla\mathbf{V} + \mathbf{U}(\nabla \cdot \mathbf{V}) - \mathbf{V}(\nabla \cdot \mathbf{U})$
7. $\nabla \times \nabla\phi = 0$
8. $\nabla \cdot \nabla \times \mathbf{U} = 0$
9. $\nabla \times \nabla \times \mathbf{V} = \nabla(\nabla \cdot \mathbf{U}) - \nabla^2\mathbf{U}$

10. $\nabla \cdot \mathbf{r} = 3$
11. $\nabla \times \mathbf{r} = 0$

PROBLEMS

1.1. Find the sum of vectors **A**, **B**, and **C** if

$$\mathbf{A} = 3i - 4j + 3k$$
$$\mathbf{B} = 2i + 4j - 6k$$
$$\mathbf{C} = -2i + 3j - 2k$$

1.2. Find the resultant $\mathbf{A} - \mathbf{B}$ if

$$\mathbf{A} = 6i + 2k \qquad \text{and} \qquad \mathbf{B} = 2i - 3j + 4k$$

1.3. Find $\mathbf{A} \times \mathbf{B}$ if

$$\mathbf{A} = -2i - 3j + 4k$$
$$\mathbf{B} = 3i + 4j - 4k$$

1.4. Find $\mathbf{A} \cdot \mathbf{B}$ if

$$\mathbf{A} = -2i - 3j + 4k$$
$$\mathbf{B} = 3i + 4j - 4k$$

1.5. Discuss vectors **M** and **N** if

$$\mathbf{M} = 2i + 3j - 4k$$
$$\mathbf{N} = 7i + 10.5j - 14k$$

1.6. Find the angle between **A** and **B** if

$$\mathbf{A} = -2i - 3j + 4k$$
$$\mathbf{B} = 3i + 4j - 4k$$

1.7. Find the angles $\mathbf{A} = 3i + 2j - 4k$ makes with the three axes.
1.8. Find the angle between the vectors **P** and **Q** if

$$\mathbf{P} = 2i + 3j - 4k$$
$$\mathbf{Q} = 3i - 2j + 6k$$

1.9. Find the vector $\mathbf{A} \times \mathbf{B}$ if

$$\mathbf{A} = 2i + 4j - 3k$$
$$\mathbf{B} = 6i - 4j + k$$

1.10. Find $\mathbf{P} \cdot \mathbf{Q}$ for the vectors of Prob. 1.8.
1.11. Find $\mathbf{B} \times \mathbf{A}$ for the vectors of Prob. 1.9.
1.12. Write out $\mathbf{V}(\nabla \cdot \mathbf{A})$
1.13. Show that, if $\mathbf{r} = ix + jy + kz$, $\nabla \cdot \mathbf{r} = 3$.
1.14. Find the angle $\mathbf{P} = 2i + 3j - 2k$ makes with the x axis.

REFERENCES

1.1. H. B. Phillips, "Vector Analysis," John Wiley & Sons, Inc., New York, 1933.

CHAPTER 2

COMPLEX VARIABLES AND FOURIER SERIES

2.1. Complex Variables. It is possible to write a vector that exists only in the xy plane as

$$z = x + iy \qquad (2.1)$$

where z is called a *complex variable.* Under these conditions x is a "real" quantity, iy an "imaginary" quantity,* and $i = \sqrt{-1}$. (This i has absolutely no connection with the unit vector i used in vector notation to indicate direction along the x axis.) The physical significance of Eq. (2.1) is shown in Fig. 2.1, where it may be seen that z is the vector from the origin to a point P and the real and imaginary quantities plot along the abscissa and ordinate, respectively.

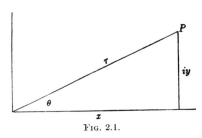

Fig. 2.1.

If we define the length of z as r, and the angle z makes with the positive x axis as θ, then

$$z = r \cos \theta + ir \sin \theta$$
$$= r (\cos \theta + i \sin \theta) \qquad (2.2)$$

Through the use of the calculus we may replace the quantity in the parentheses by a shorter term. For instance, let

$$m = \cos \theta + i \sin \theta$$

Then, differentiating,

$$\frac{dm}{d\theta} = - \sin \theta + i \cos \theta = i(\cos \theta + i \sin \theta)$$

or

$$\frac{dm}{m} = i\, d\theta$$

Integrating and noting that the constant of integration is zero, we get

$$m = e^{i\theta} = \cos \theta + i \sin \theta \qquad (2.3)$$

* It is regrettable that this particular term was selected, since imaginary quantities are no more unreal than any others.

12

so that Eq. (2.2) may evidently be written

$$z = re^{i\theta} \tag{2.4}$$

The term r is called the *modulus* of z ($= \text{mod } z$) and θ the *argument* of z ($= \text{arg } z$).

The use of complex variables enhances the possibility of mathematical manipulation since the rules of algebra are readily applicable. Thus if $z_1 = r_1 e^{i\theta_1}$ and $z_2 = r_2 e^{i\theta_2}$,

$$z_1 z_2 = r_1 r_2 e^{i(\theta_1 + \theta_2)} \tag{2.5}$$

$$\frac{z_1}{z_2} = \frac{r_1}{r_2} e^{i(\theta_1 - \theta_2)}$$

The logarithm of a complex variable comes directly from Eq. (2.4) as

$$\ln z = \ln re^{i\theta} = \ln r + i\theta$$

The conjugate of z, written $\bar{z}$, is defined as

$$\bar{z} = x - iy$$

It has the same absolute value as z. From the above,

$$\left.\begin{array}{c} z\bar{z} = x^2 + y^2 \\ z - \bar{z} = 2iy \\ z + \bar{z} = 2x \end{array}\right\} \tag{2.6}$$

Equation (2.5) above leads to the connection between i and the vertical axis. When $\theta = \pi/2$, $e^{i\theta} = \cos \theta + i \sin \theta = i$. Since multiplying a vector by $e^{i\theta}$ rotates it through $\theta°$ to the left, it is seen that multiplying a vector by i ($= e^{i\theta}$ when $\theta = \pi/2$) rotates it 90° to the left. Thus $z = x + iy$ is an orthogonal representation. It may also be shown that multiplying a vector (or complex variable) by i^2 ($= -1$) rotates it 180° to the left.

2.2. Properties of Complex Variables. Since we have put $z = x + iy$, consider a function

$$w = f(z)$$

Note that, if x or y changes, z will change and also w. But the converse is not exactly true; a change in w requires a change in z, but not necessarily a change in x and y. It may be inferred that the partial derivatives denoting change produced by varying one variable only will be of importance.

Let us assign some value to w in terms of z, say

$$w = z^2 \tag{2.7}$$

Then

$$w = (x + iy)^2 = x^2 + 2ixy - y^2$$

We see that there is a real *and* an imaginary part, and it develops that the two always appear together. Let us call the real part ϕ and the imaginary part ψ. Then

$$w = \phi + i\psi \tag{2.8}$$

Interesting meanings will be attached to the terms of Eq. (2.8) later on.

2.3. Properties of ϕ and ψ. From the relation

$$z = x + iy$$

we have

$$\frac{\partial z}{\partial x} = 1 \qquad \frac{\partial x}{\partial z} = 1$$

$$\frac{\partial z}{\partial y} = i \qquad \frac{\partial y}{\partial z} = -i \tag{2.9}$$

Now, if a quantity is a function of one variable, the partial derivative with respect to that variable is the same as the total derivative. Hence, as w varies with z,

$$\frac{dw}{dz} = \frac{\partial w}{\partial z} = \frac{\partial w}{\partial x}\frac{\partial x}{\partial z} = \left(\frac{\partial \phi}{\partial x} + i\frac{\partial \psi}{\partial x}\right) \cdot 1 \tag{2.10}$$

or

$$\frac{dw}{dz} = \frac{\partial w}{\partial z} = \frac{\partial w}{\partial y}\frac{\partial y}{\partial z} = \left(\frac{\partial \phi}{\partial y} + i\frac{\partial \psi}{\partial y}\right) \cdot (-i)$$

$$= -i\frac{\partial \phi}{\partial y} + \frac{\partial \psi}{\partial y} \tag{2.11}$$

or, shorter,

$$\frac{dw}{dz} = \frac{\partial w}{\partial z} = \frac{\partial w}{\partial x} \tag{2.12}$$

and

$$\frac{dw}{dz} = \frac{\partial w}{\partial z} = \frac{1}{i}\frac{\partial w}{\partial y} \tag{2.13}$$

It is perhaps in order to make a few comments on Eqs. (2.10) through (2.13) in order to explain a seemingly paradoxical situation. To begin with, in complex variables the student must divorce the usual concept of dw/dz being a "slope" and think of it as the rate of change of w with z. Now, if we plot values of w on ϕ, $i\psi$ axes and z on x, iy axes, since w is a function of z for each point w there will be a corresponding point z. (In this work only continuous functions are of importance.) At point z, the rate of change of w with z has some particular value, and of course it has that same value regardless of whether we approach z from left or right, up or down.

Thus since by definition

$$\frac{dw}{dz} = \lim_{\Delta z \to 0} \frac{\Delta w}{\Delta z} = \lim_{\Delta z \to 0} \frac{f(z_0 + \Delta z) - f(z_0)}{\Delta z}$$

The equation

$$\frac{dw}{dz} = \frac{\partial w}{\partial z} = \frac{\partial w}{\partial x} = \frac{\partial w}{\partial(iy)}$$

merely states that the rate of change of w with z is the same regardless of the manner in which it is approached.

Equating reals and imaginaries in Eqs. (2.10) and (2.11), we have

$$\frac{\partial \phi}{\partial x} = \frac{\partial \psi}{\partial y} \tag{2.14}$$

and

$$\frac{\partial \phi}{\partial y} = -\frac{\partial \psi}{\partial x} \tag{2.15}$$

Equations (2.14) and (2.15) are used frequently in later work. Additional equations may be formed by manipulating the above results. For instance, since

$$\frac{dw}{dz} = \frac{\partial \phi}{\partial x} + i\frac{\partial \psi}{\partial x} \tag{2.10}$$

and since

$$\frac{\partial \psi}{\partial x} = -\frac{\partial \phi}{\partial y} \tag{2.15}$$

we have

$$\frac{dw}{dz} = \frac{\partial \phi}{\partial x} - i\frac{\partial \phi}{\partial y} \tag{2.16}$$

In a similar manner

$$\frac{dw}{dz} = \frac{\partial \psi}{\partial y} + i\frac{\partial \psi}{\partial x} \tag{2.17}$$

2.4. Cauchy's Second Theorem. The laws of complex variables may be applied to the integration of series in a most useful manner. Consider the function represented by

$$f(z) = \int_C z^n \, dz$$

where z is a complex variable and n is an integer, and consider the region R with a contour C. It will be seen later that we may deform the contour C into a circle C' without changing the value of the line integral (see Sec. 4.2). After we have performed this change, $z = re^{i\theta}$ differentiates to

$$\frac{dz}{d\theta} = ire^{i\theta}$$

since $\dfrac{dr}{d\theta} = 0$ for a circle.

Then the integral

$$\int_C z^n\, dz = \int_{C'} z^n\, dz$$

and

$$\int_C z\, dz = \int_{C'} re^{i\theta} \cdot ire^{i\theta}\, d\theta$$
$$= \int_{C'} ir^2 e^{2i\theta}\, d\theta$$

or, in general form,

$$\int_C z^n\, dz = ir^{n+1} \int_{C'} e^{i(n+1)\theta}\, d\theta$$

and

$$\int_C z^n\, dz = \int_0^{2\pi} ir^{n+1} e^{i(n+1)\theta}\, d\theta = ir^{(n+1)} \frac{e^{i(n+1)\theta}}{i(n+1)} \Big]_0^{2\pi}$$
$$= ir^{n+1} \left[\frac{e^{2\pi i(n+1)} - e^0}{i(n+1)} \right]$$

This expression vanishes for all values of n except $n = -1$, at which time it becomes indeterminate, but easily evaluated as follows: If $n = -1$

$$\int_C z^n\, dz = \int_{C'} \frac{dz}{z} = \int_0^{2\pi} \frac{ire^{i\theta}\, d\theta}{re^{i\theta}}$$
$$= \int_0^{2\pi} i\theta\, d\theta = 2\pi i$$

We therefore conclude that the integral of a complex series taken about a closed contour depends only on the term having $n = -1$, or

$$\int_C \sum_{-\infty}^{\infty} A_n z^n\, dz = \int_C (\cdots A_2 z^2 + A_1 z + A_0 + A_{-1} z^{-1} + A_{-2} z^{-2} \cdots)\, dz$$
$$= 2\pi i A_{-1}$$

This expression, known as *Cauchy's second theorem*, is most useful in evaluating series integrations that appear later.

2.5. The Laplacian. From Eqs. (2.14) and (2.15)

$$\frac{\partial\phi}{\partial x} = \frac{\partial\psi}{\partial y}; \qquad \therefore \frac{\partial^2\phi}{\partial x^2} = \frac{\partial^2\psi}{\partial x\,\partial y} \tag{2.18}$$

$$\frac{\partial\phi}{\partial y} = -\frac{\partial\psi}{\partial x}; \qquad \therefore \frac{\partial^2\phi}{\partial y^2} = -\frac{\partial^2\psi}{\partial x\,\partial y} \tag{2.19}$$

Adding, we get

$$\frac{\partial^2\phi}{\partial x^2} + \frac{\partial^2\phi}{\partial y^2} = 0 \qquad (2.20)$$

And in a similar manner

$$\frac{\partial^2\psi}{\partial x^2} + \frac{\partial^2\psi}{\partial y^2} = 0 \qquad (2.21)$$

Equations (2.20) and (2.21) are called the Laplacians of ϕ and ψ and may be written $\nabla^2\phi$ and $\nabla^2\psi$. It will be shown later that when $\nabla^2\phi = 0$ the flow represented by ϕ is incompressible and when $\nabla^2\psi = 0$ the flow represented by ψ is irrotational, these terms to be later discussed and defined.

2.6. Matching Empirical Curves by Simultaneous Power Equations. Many important developments in aerodynamics have been empirical in nature, and as theories were advanced, it became necessary to have mathematical equations for the empirical results. Examples of this nature include the equation of the basic NACA four-digit symmetrical shape, thin-airfoil theory, and span loading conditions. The mathematician has two tools particularly well suited for attacking these equation problems—power and Fourier's series.

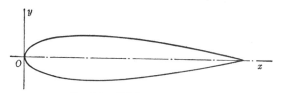

Fig. 2.2.   NACA 0020 profile.

Let us consider the development of the equation of the NACA 0020 airfoil. During the era 1925 to 1935 two airfoils of widely different appearance were quite "popular"—the Göttingen 398 and the Clark Y. Upon removal of the camber both of these airfoils were found to have a nearly identical thickness distribution, and it was decided to use this thickness distribution with a number of parabolic cambers to determine the effect of maximum camber location and amount. An important step became the determination of the equation of the basic shape. The empirical dimensions available were for a 20 per cent thick airfoil emplaced on the xy axes with the leading edge at the origin and the trailing edge at $x = 1.0$:

1. $y = 0.1$ when $x = 0.3$
2. Slope $= 0$ when $x = 0.3$
3. Thickness at trailing edge $= 0.002$ when $x = 1.0$

4. Slope at trailing edge $= -0.234$ when $x = 1.0$
5. The nose shape yielded $y = 0.078$ when $x = 0.1$

Looking at the airfoil shown in Fig. 2.2, we are immediately struck with the possibility that the nose shape approximates a parabola ($y = a_0 \sqrt{x}$, where a_0 is a constant). However, we must "close" the parabola at the

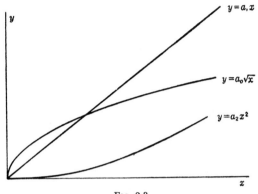

FIG. 2.3.

trailing edge, and it seems reasonable to subtract a straight line ($y = a_1 x$) as a first approximation. We now have

$$y = a_0 \sqrt{x} + a_1 x \tag{2.22}$$

In addition we have five empirical conditions to meet. The method is to add them in as additional terms to our general equation (2.22), getting

$$y = a_0 \sqrt{x} + a_1 x + a_2 x^2 + a_3 x^3 + a_4 x^4 \tag{2.23}$$

Differentiation yields

$$\frac{dy}{dx} = 0.50 a_0 x^{-\frac{1}{2}} + a_1 + 2x a_2 + 3x^2 a_3 + 4x^3 a_4 \tag{2.24}$$

We now substitute our data into Eq. (2.23) or (2.24) as needed, obtaining for the five conditions

1. $0.548 a_0 + 0.3 a_1 + 0.09 a_2 + 0.027 a_3 + 0.0081 a_4 = 0.1$
2. $0.913 a_0 + a_1 + 0.6 a_2 + 0.27 a_3 + 0.108 a_4 = 0$
3. $a_0 + a_1 + a_2 + a_3 + a_4 = 0.002$
4. $0.5 a_0 + a_1 + 2a_2 + 3a_3 + 4a_4 = -0.234$
5. $0.316 a_0 + 0.1 a_1 + 0.01 a_2 + 0.001 a_3 + 0.0001 a_4 = 0.078$

Having five equations for five unknowns, the simultaneous equations may be solved. They then yield

$$y = 0.29690 \sqrt{x} - 0.12600 x - 0.35160 x^2 + 0.28430 x^3 - 0.10150 x^4 \tag{2.25}$$

which is the equation of the airfoil. To be more precise, it is the equation of one line through the points specified; other lines through the same points may of course be drawn. This leads us to note that, strictly speaking, Eq. (2.25) should not be expected to yield exact slopes at points other than those used in its derivation, but in this particular case sufficient accuracy is actually obtained.

In passing, it was not necessary to have a $\sqrt{x}$ term in Eq. (2.23) at all; but its exclusion would require that many more points be used in the derivation. It is quite reasonable that any curve matching should start with the best possible assumptions in order to simplify the calculations.

2.7. Matching Empirical Curves by Simultaneous Fourier Series. A second approach employing sine and cosine series is often more expeditious than using exponential relations. Two well-known examples of this type appear in the aerodynamic theory. The first series is needed to represent a camber line not symmetrical about its mid-point and the second is needed to represent a distribution of lift across a wing, symmetrical about its mid-point. Considering the second problem first, we write

$$y = -s \cos \theta \qquad (2.26)$$

where s = wing semi-span.

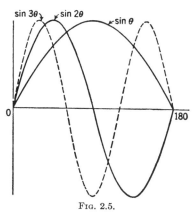

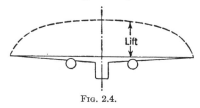

Fig. 2.4.   Fig. 2.5.

Hence θ varies from 0 to π as we travel from port to starboard. Now consider the series (plotted in Fig. 2.4)

$$L = \sum_{1}^{\infty} A_n \sin n\theta \qquad (2.27)$$

Sin θ (see Fig. 2.5) is symmetrical about its mid-point from 0 to π, but sin 2θ is not.

Sin 3θ becomes symmetrical again, and it can hence be shown that a series sin $n\theta$ will remain symmetrical about $\theta = 90°$ as long as n is odd. In a manner similar to the exponential approach we write the general equation

$$L = A_1 \sin \theta + A_3 \sin 3\theta + A_5 \sin 5\theta \qquad (2.28)$$

and substitute known points and slopes in the equation as before.

Suppose, for example, it is desired to write the equation of the solid line in Fig. 2.6. We have $L = 1.0$, $\theta = 90°$; $L = 0.8$, $\theta = 60°$; $L = 0.5$, $\theta = 30°$. Substituting in Eq. (2.28)

$$\left.\begin{array}{l} A_1 \sin 90 + A_3 \sin 270 + A_5 \sin 450 = 1.0 \\ A_1 \sin 60 + A_3 \sin 180 + A_5 \sin 300 = 0.8 \\ A_1 \sin 30 + A_3 \sin 90 + A_5 \sin 150 = 0.5 \end{array}\right\} \qquad (2.29)$$

from which

$$A_1 = 0.962$$
$$A_3 = 0$$
$$A_5 = 0.038$$

and the equation of the line becomes

$$L = 0.962 \sin \theta + 0.038 \sin 5\theta \qquad (2.30)$$

Of interest is the development of $A_3 = 0$ as it shows that the contribution of some terms in the series may be negligible. Later series converge so rapidly that we frequently need but two or three terms.

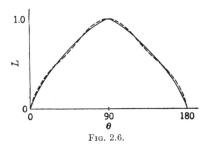

Fig. 2.6.

The manner in which Eq. (2.30) simulates the conditions is shown in Fig. 2.6. It is evident that more points should have been used. The first example mentioned above, the camber line not symmetrical about its mid-point, can use a series including both odd and even values of n.

2.8. Hyperbolic Functions. In Chap. 8 some of the mathematical steps are considerably shortened by the use of hyperbolic functions. The presentation is most elementary, however, and the student will be adequately prepared by recalling the fundamentals of hyperbolic functions as follows:

1. By definition

$$\left.\begin{array}{l} \sinh x = \dfrac{e^x - e^{-x}}{2} \\[2mm] \cosh x = \dfrac{e^x + e^{-x}}{2} \end{array}\right\} \qquad (2.31)$$

2. Equation (2.31) may be developed to yield

$$\cosh^2 x - \sinh^2 x = 1 \qquad (2.32)$$

3. In a manner similar to the trigonometric functions

$$\sinh x \cong x \quad \text{and} \quad \cosh x \cong 1 \quad \text{for small values of } x$$

4. $\cos ix = \cosh x$
   $\sin ix = i \sinh x$
   $\tan ix = i \tanh x$ $\quad\quad$ (2.33)

5. $\sin x = \dfrac{e^{ix} - e^{-ix}}{2i}$

   $\cos x = \dfrac{e^{ix} + e^{-ix}}{2}$ $\quad\quad$ (2.34)

The last two relations come from the development of Eq. (2.4).

2.9. Determinants. The need for solving simultaneous equations arises several times in later chapters, and since they are sometimes of the fourth order, it appears advisable to state a shortened form which is a great timesaver.

First of all, the solutions for simultaneous equations may be written directly using determinants according to the rule that if

$$a_1x + b_1y + c_1z = k$$
$$a_2x + b_2y + c_2z = l$$
$$a_3x + b_3y + c_3z = m$$

then

$$x = \frac{\begin{vmatrix} k & b_1 & c_1 \\ l & b_2 & c_2 \\ m & b_3 & c_3 \end{vmatrix}}{\begin{vmatrix} a_1 & b_1 & c_1 \\ a_2 & b_2 & c_2 \\ a_3 & b_3 & c_3 \end{vmatrix}} ; \; y = \frac{\begin{vmatrix} a_1 & k & c_1 \\ a_2 & l & c_2 \\ a_3 & m & c_3 \end{vmatrix}}{\begin{vmatrix} a_1 & b_1 & c_1 \\ a_2 & b_2 & c_2 \\ a_3 & b_3 & c_3 \end{vmatrix}} ; \text{ and } z = \frac{\begin{vmatrix} a_1 & b_1 & k \\ a_2 & b_2 & l \\ a_3 & b_3 & m \end{vmatrix}}{\begin{vmatrix} a_1 & b_1 & c_1 \\ a_2 & b_2 & c_2 \\ a_3 & b_3 & c_3 \end{vmatrix}} \quad (2.35)$$

Simple cross multiplication is adequate for third-order determinants such as those above, but fourth-order determinants cannot be so treated, and by many methods their solution is a long and laborious task. The simplest system for fourth-order solution that has come to the attention of the author is that of Laplace's development by columns, using complementary minors. Without going into the theory, a fourth-order determinant may be evaluated by the relation

$$\begin{vmatrix} a_1 & b_1 & c_1 & d_1 \\ a_2 & b_2 & c_2 & d_2 \\ a_3 & b_3 & c_3 & d_3 \\ a_4 & b_4 & c_4 & d_4 \end{vmatrix} = \begin{vmatrix} a_1 & b_1 \\ a_2 & b_2 \end{vmatrix} \cdot \begin{vmatrix} c_3 & d_3 \\ c_4 & d_4 \end{vmatrix} - \begin{vmatrix} a_1 & b_1 \\ a_3 & b_3 \end{vmatrix} \cdot \begin{vmatrix} c_2 & d_2 \\ c_4 & d_4 \end{vmatrix}$$
$$+ \begin{vmatrix} a_1 & b_1 \\ a_4 & b_4 \end{vmatrix} \cdot \begin{vmatrix} c_2 & d_2 \\ c_3 & d_3 \end{vmatrix} + \begin{vmatrix} a_2 & b_2 \\ a_3 & b_3 \end{vmatrix} \cdot \begin{vmatrix} c_1 & d_1 \\ c_4 & d_4 \end{vmatrix}$$
$$- \begin{vmatrix} a_2 & b_2 \\ a_4 & b_4 \end{vmatrix} \cdot \begin{vmatrix} c_1 & d_1 \\ c_3 & d_3 \end{vmatrix} + \begin{vmatrix} a_3 & b_3 \\ a_4 & b_4 \end{vmatrix} \cdot \begin{vmatrix} c_1 & d_1 \\ c_2 & d_2 \end{vmatrix} \quad (2.36)$$

2.10. The Binomial Expansion. Many of the expressions used in the theory to follow require the use of the binomial theorem in their derivation, and it is hence listed for convenience.

$$(a \pm b)^n = a \pm \frac{na^{(n-1)}}{1!} b + \frac{n(n-1)}{2!} a^{(n-2)}b^2$$
$$\pm \frac{n(n-1)(n-2)}{3!} a^{(n-3)}b^3 \cdots \quad (2.37)$$

2.11. Axes. Perhaps, before proceeding further, a discussion on the various methods used to locate a point relative to fixed axes may be in order. It has been customary in lower level mathematics courses to locate a point according to an abscissa and an ordinate, (x, y) for example. A second method is called *polar representation* and points located by this

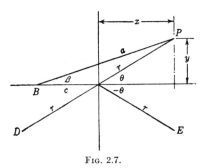

Fig. 2.7.

system are determined by a radius and an angle to a reference as (r, θ). Occasionally, y and θ may be used for convenience. It is apparent that these two parameters will locate one point and only one point just as do the others. Indeed, y may be measured vertically and θ out along the horizontal axis.

In complex variables the radius and angle may be combined into a single symbol (such as z) which may then be returned to rectilinear coordinates ($z = x + iy$) or polar coordinates ($z = re^{i\theta}$). Sometimes one finds the complex axes labeled the x and iy axes.

To illustrate the systems outlined above, consider Fig. 2.7. The point P is at (x, y) in rectilinear coordinates; at (r, θ) in polar coordinates; or at (y, θ) in the mixed-coordinates system. In the complex plane (as it is called) we have $z_P = re^{i\theta} = x + iy$. To illustrate the use of the minus sign for complex variables we have $z_D = -z_P = -re^{i\theta}$ and $z_E = re^{-i\theta}$. The location of B may be written as $z_B = -ce^{i0} = ce^{i\pi} = -c$ or as the sum of two complex variables $z_B = re^{i\theta} - ae^{i\beta}$.

PROBLEMS

2.1. Prove that $e^{-i\theta} = \cos\theta - i\sin\theta$.

2.2. If in vector notation $z = 3i + 2j$, write the conjugate $\bar{z}$.

2.3. From the exponential expressions for $\sinh x$ and $\cosh x$ derive the expression for $\tanh x$.

2.4. Show how multiplying a complex variable by $-i$ rotates the variable 90° to the right.

2.5. If $w = \phi + i\psi$ and $z = x + iy$, prove that $\dfrac{dw}{dz} = \dfrac{\partial w}{\partial x}$.

REFERENCES

2.1. W. F. Durand, "Aerodynamic Theory," Vol. I, pp. 1–21, Springer, Berlin, 1935.

CHAPTER 3

THE STREAM FUNCTION

The first task in making an attack on aerodynamic theory is the development of mathematical tools by which we may express the flow about various bodies and the laws that govern such flows.

Fundamentally we have two rules: the flow must obey Newton's second law at all times, and the mass entering a zone must equal that leaving it (this implies steady motion such that the partials with respect to time vanish) unless the zone contains "singular points" (see Sect. 3.10). In addition, certain assumptions and restrictions may be considered to simplify expressions that would otherwise become excessively lengthy.

Our starting point is perhaps some work presented by Sir Isaac Newton (1642–1727) (Ref. 3.1) in 1687 wherein mention is made of a "rare medium" that possessed no frictional qualities. With this assumption Newton laid the foundation for very simple and useful flow relations.

3.1. Fluids: Perfect Fluids. A fluid is a homogeneous material that deforms under very small shear stress. Thus we may consider air a fluid, and indeed the science of fluid dynamics becomes simplified (and known as *aerodynamics*) when the fluid under consideration has low density, so that gravitational effects are negligible, and low viscosity, so that shear forces (skin friction forces) are not preponderant.

Air at sea level is of course homogeneous, compressible, and slightly viscous. But the amount of compressibility encountered as long as all velocities remain low is quite insignificant, and the effect of viscosity, if we consider flow outside the boundary layer, is also small—with several important exceptions to be considered later. With these reservations we gladly accept Newton's rare medium, calling it a *perfect incompressible fluid*, and assuming that it has very small viscosity.* There will be occasions where both of these assumptions will invite further investigation.

In the half century following Newton three great contributors stand out: Daniel Bernoulli (1700–1783), d'Alembert (1717–1783), and Euler

---

* Zero viscosity would, of course, correspond to infinite Reynolds number where Reynolds number is defined as $RN = (\rho/\mu)Vl$ (ρ = air density, slug per cu ft, μ = viscosity, lb-sec per sq ft, V = velocity, ft per sec, and l = characteristic length, ft) after the work of Reynolds (1842–1912).

(1707–1783). Each made contributions that we need to examine more fully. First, however, a note on our approach.

There are, in classical fluid dynamics, two approaches to the solution of problems concerning bodies in fluids. The first arrangement, and by far the most common in practical aerodynamics, is that which occurs when the *body moves* through a still fluid. The second is that which arises when the *body is still* and the fluid moves. While the practical example of this relation is confined to the wind tunnel and water channel, the theoretical treatment of body-still-fluid-moving is so far the simpler that it is universally employed and will be continued in these pages. The principle of relativity tells us that the forces and moments on a body are due to relative motion, and no extraneous qualities arise through fixing the body.

3.2. The Equation of Continuity (Conservation of Mass). D'Alembert around 1744 observed that, if one specified a fixed zone in a moving fluid, the mass of fluid entering that zone must equal the mass leaving it, provided that the flow was *continuous* (hence the name) and the zone contained no singular points. Useful fluid mechanics equations may be derived as follows (see Fig. 3.1): Consider a point in space P, and let the velocities at that point be u, v, and w along their respective axes and

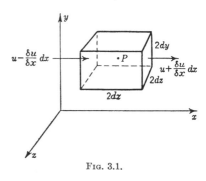

Fig. 3.1.

the pressure and density there be p and ρ, respectively. Around P erect a zone extending along each axis a distance dx, dy, and dz. The volume of the zone will then be $8dx\,dy\,dz$. Since the rate of change in horizontal velocity along the x axis is $\dfrac{\partial u}{\partial x}$, the average velocity entering the face $ABCD$ is $u - \dfrac{\partial u}{\partial x}\,dx$, and that leaving the face $DEFG$ is $u + \dfrac{\partial u}{\partial x}\,dx$. The respective masses for the two faces are $\left(\rho - \dfrac{\partial \rho}{\partial x}\,dx\right)\left(u - \dfrac{\partial u}{\partial x}\,dx\right)4dy\,dz$ and $\left(\rho + \dfrac{\partial \rho}{\partial x}\,dx\right)\left(u + \dfrac{\partial u}{\partial x}\,dx\right)4dy\,dz$. Along the y axis the mass entering is $\left(\rho - \dfrac{\partial \rho}{\partial y}\,dy\right)\left(v - \dfrac{\partial v}{\partial y}\,dy\right)4dx\,dz$, and that leaving is $\left(\rho + \dfrac{\partial \rho}{\partial y}\,dy\right)\left(v + \dfrac{\partial v}{\partial y}\,dy\right)4dx\,dz$. Writing a similar relation for the z axis, and applying the principle that the mass entering must equal the mass leaving, we have (after simplification and dividing through by the volume of the zone) the continuity equation

$$\rho \left(\frac{\partial u}{\partial x} + \frac{\partial v}{\partial y} + \frac{\partial w}{\partial z} \right) + u \frac{\partial \rho}{\partial x} + v \frac{\partial \rho}{\partial y} + w \frac{\partial \rho}{\partial z} = 0$$

For the fluid to be incompressible, the partials of ρ must be zero. We then have a condition that motion of a continuous incompressible fluid must satisfy

$$\frac{\partial u}{\partial x} + \frac{\partial v}{\partial y} + \frac{\partial w}{\partial z} = 0 \tag{3.1}$$

or, in vector notation,

$$\nabla \cdot \mathbf{q} = 0 \tag{3.2}$$

The quantity $\nabla \cdot \mathbf{q}$ is called the *divergence of q*.

The use of Eqs. (3.1) and (3.2) will be discussed more fully in Chapter 4.

3.3. Euler's Equations of Motion. D'Alembert also postulated the principle of conservation of momentum, and Euler, presenting a similar principle in more complete form, developed the dynamic equations of motion which today bear his name.

The equations of motion for continuous flow (all partials with respect to time vanish) may be derived as follows: Consider a zone as shown in Fig. 3.1. The average pressure force over the face *ABCD* will be $\left(p - \frac{\partial p}{\partial x} dx \right) 4dy\, dz$, and that over the face *EFGH* will be $\left(p + \frac{\partial p}{\partial x} dx \right)$ $4dy\, dz$ so that the net pressure force on the fluid will be $- \frac{\partial p}{\partial x} 8dx\, dy\, dz$.

Now suppose a body exists at P such that it exerts a force X per unit mass. Its total force will be $8X\rho\, dx\, dy\, dz$.

Writing $F = ma$ for the fluid in the zone, we have

$$8\rho X\, dx\, dy\, dz - 8 \frac{\partial p}{\partial x} dx\, dy\, dz = 8\rho\, dx\, dy\, dz\, a_x$$

where a_x = average acceleration of the fluid in the zone in the x direction.

Dividing through by the mass of the fluid in the zone, we have

$$X - \frac{1}{\rho} \frac{\partial p}{\partial x} = a_x$$

and for the other axes

$$Y - \frac{1}{\rho} \frac{\partial p}{\partial y} = a_y$$

$$Z - \frac{1}{\rho} \frac{\partial p}{\partial z} = a_z$$

Now u is a function of x, y, and z, so the total derivative of u may be written

$$du = \frac{\partial u}{\partial x}\,dx + \frac{\partial u}{\partial y}\,dy + \frac{\partial u}{\partial z}\,dz$$

and, dividing by dt (a short time increment), we have

$$\frac{du}{dt} = a_x = u\,\frac{\partial u}{\partial x} + v\,\frac{\partial u}{\partial y} + w\,\frac{\partial u}{\partial z} \tag{3.3}$$

and for the other acceleration components

$$a_v = u\,\frac{\partial v}{\partial x} + v\,\frac{\partial v}{\partial y} + w\,\frac{\partial w}{\partial z}$$

$$a_z = u\,\frac{\partial w}{\partial x} + v\,\frac{\partial w}{\partial y} + w\,\frac{\partial w}{\partial z}$$

By substitution the final equalities of motion for continuous flow become

$$X - \frac{1}{\rho}\frac{dp}{dx} = u\,\frac{\partial u}{\partial x} + v\,\frac{\partial u}{\partial y} + w\,\frac{\partial u}{\partial z} \tag{3.4}$$

$$Y - \frac{1}{\rho}\frac{dp}{dy} = u\,\frac{\partial v}{\partial x} + v\,\frac{\partial v}{\partial y} + w\,\frac{\partial v}{\partial z} \tag{3.5}$$

$$Z - \frac{1}{\rho}\frac{dp}{dz} = u\,\frac{\partial w}{\partial x} + v\,\frac{\partial w}{\partial y} + w\,\frac{\partial w}{\partial z} \tag{3.6}$$

Their use will be demonstrated at a later time.

3.4. Bernoulli's Equation. The particular equation that carries Bernoulli's name was presented by him in 1738. Briefly, it states that a fluid has two kinds of pressure, static and dynamic, and their sum (the total pressure) remains constant along any stream filament within limits to be discussed later. The Bernoulli equation may be derived from Euler's equations or in a number of other ways. Perhaps the clearest is the simple equality of work and change in kinetic energy as follows:

Consider an incompressible airstream moving to the right (Fig. 3.2) with increasing velocity, and suppose $ABCD$ represents a

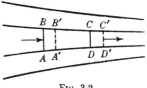

Fig. 3.2

portion of a stream tube (a zone bounded by streamlines—see Sect. 3.6). After a short time t the mass of air will be at $A'B'C'D'$, and the mass

$$ABB'A' = CDD'C'$$

since $A'B'CD$ is common.

Let p_1, v_1, S_1 be the average pressure, velocity, and area of volume $ABB'A'$, and p_2, v_2, S_2 be the corresponding quantities of $CDD'C'$. Then the volume

$$ABB'A' = S_1v_1t$$

and

$$CDD'C' = S_2v_2t$$

Hence $S_1v_1t = S_2v_2t = k$, and we see that streamlines draw together as the local velocity increases, provided that the flow is incompressible.

Now the effect of the contracting stream tube has been to replace a mass $ABB'A'$ at velocity v_1 by an equal mass $CDD'C'$ at velocity v_2. The net change in kinetic energy is

$$\Delta\text{K.E.} = \tfrac{1}{2}\rho k(v_2{}^2 - v_1{}^2)$$

The only forces acting on the motion are the pressure forces p_1S_1 to the right and p_2S_2 to the left, acting through the distances v_1t and v_2t, respectively. The total work done by the pressure forces is then

$$p_1S_1v_1t - p_2S_2v_2t = (p_1 - p_2)k \tag{3.7}$$

Equating the change in kinetic energy to the work done we have

$$\tfrac{1}{2}\rho k(v_2{}^2 - v_1{}^2) = (p_1 - p_2)k$$

or

$$p_1 + \tfrac{1}{2}\rho v_1{}^2 = p_2 + \tfrac{1}{2}\rho v_2{}^2 = \text{const} \tag{3.8}$$

Equation (3.8) is Bernoulli's equation for an incompressible fluid. Defining

$$C_p = \frac{p - p_0}{(\rho/2)V^2} \tag{3.9}$$

and writing Eq. (3.9) between a point on a still body and one in the free stream far removed from the body, we get

$$C_p = 1 - \left(\frac{v}{V_0}\right)^2 \tag{3.10}$$

where v = local velocity, ft per sec
  V_0 = free-stream velocity, ft per sec
Equation (3.10) is useful in airfoil work.

3.5. Boundary Conditions. In the development of Euler's equations of motion the case of a solid body being submerged in a moving fluid was mentioned. At a point on the body two conditions must prevail: (1) no normal velocity can exist; (2) the local angle of flow must equal the local inclination of the body surface. These *boundary* conditions will be found useful in later work.

3.6. The Stream Function and Streamlines. Having now some basic laws for fluid flow, we turn to a mathematical and graphical method of presenting and studying flow patterns which follows suggestions made by Rankine (1820–1872) in 1864. It consists in plotting lines—called *stream-lines*—which are everywhere tangent to the fluid velocity vectors, and noting that where they crowd together the static pressure is reduced. Where they tend to part, the local static pressure is increased. The bodies we shall analyze in this chapter have no superior qualities as far as having low drag is concerned, especially since a perfect fluid solution with zero drag can be found for all two-dimensional bodies.* They simply represent the simplest shapes for which we can write the flow equations, and they lead to shapes whose practical value is unmistakable.

Starting at this point and continuing through Chap. 8, we shall be concerned with "two-dimensional" flow only. For this type (flow in the xy plane) only the u and v velocities exist. The velocity w and all partials with respect to z are zero. For convenience we shall on occasion assume a unit (1 ft) depth. This assumption enables us to talk of flow quantity in the usual cubic feet per second units instead of the less understandable two-dimensional quantity units, square feet per second.

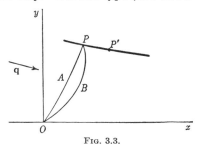

Fig. 3.3.

Now let us apply the equation of continuity to a system of streamlines. Consider Fig. 3.3, and let ψ (psi) be the volume of fluid that flows between a point P and the origin. Identically, ψ—called the *stream function*—is the amount of fluid that crosses line OAP or OBP or, indeed, any other single-valued curve that connects the origin and point P.

Consider a second point P' so located that no flow crosses PP'. Then ψ for P will equal that for P', and the line PP' is called a *streamline:* a streamline is everywhere tangent to the flow direction. A region bounded by streamlines is a *stream channel* or *stream tube.*

3.7. ψ in Rectilinear Coordinates. Consider a velocity field in which horizontal flow u to the right and vertical flow v upward are positive, and let there be, say, a steady flow downward and to the right. Choosing two streamlines quite close together we have for their quantities ψ and

---

* Kirchhoff (Ref. 3.2 and elsewhere) has presented a perfect fluid solution for two-dimensional bodies in which a drag is developed.

$\psi + d\psi$. The amount of fluid flowing in the *stream channel* between them is $d\psi$.

Let a small control zone be emplaced as shown in Fig. 3.4. The quantity of fluid flowing into the zone will be $d\psi$, while that flowing out will be, componentwise, $u\,dy - v\,dx$. The negative sign for the vertical amount springs from the downward velocity being negative. That is,

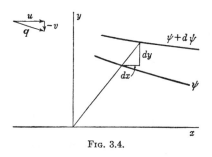

$$d\psi = u\,dy - v\,dx$$

From the definition of a derivative

$$d\psi = \frac{\partial \psi}{\partial x}\,dx + \frac{\partial \psi}{\partial y}\,dy$$

and hence, equating components,

$$u = \frac{\partial \psi}{\partial y}; \qquad v = -\frac{\partial \psi}{\partial x} \qquad (3.11)$$

Fig. 3.4.

This means that, if we have a stream function in x and y, the velocity equations may be obtained by taking the proper partial derivatives. Velocities at special points may then be obtained by substituting the coordinates of the points in the velocity formulas.

Example 3.1. Find the velocity components, total velocity, and inclination of the flow represented by the stream function $\psi = x^2 - y^2$ at the point (2, 6).

1. $u = \dfrac{\partial \psi}{\partial y} = -2y$ or $(-2)(6) = -12$ ft per sec at $y = 6$

2. $v = -\dfrac{\partial \psi}{\partial x} = -2x$ or $(-2)(2) = -4$ ft per sec at $x = 2$

3. $q = \sqrt{u^2 + v^2} = \sqrt{(-12)^2 + (-4)^2} = 12.65$ ft per sec

4. $\theta = \tan^{-1}\dfrac{v}{u} = \tan^{-1} 0.333 = 198.4°$ measured from $\theta = 0°$. The velocity is down and to the left

3.8. ψ in Polar Coordinates. The stream function in polar coordinates may be treated in a similar manner. The conventions for flow direction now embrace tangential flow in a counterclockwise manner and outward radial flow as positive (see Fig. 3.5). We now have

$$-d\psi = -q_r r\,d\theta + q_\theta\,dr$$

and

$$d\psi = \frac{\partial \psi}{\partial \theta}\,d\theta + \frac{\partial \psi}{\partial r}\,dr$$

from which

$$q_r = \frac{1}{r}\frac{\partial \psi}{\partial \theta}; \qquad q_\theta = -\frac{\partial \psi}{\partial r} \qquad (3.12)$$

This time the stream function in r and θ must be differentiated to yield the velocity components—a minor difficulty, shown in the example below, arising when the flow direction is determined.

Example 3.2. Find the velocity components, total velocity, and inclination of the flow represented by $\psi = r\theta^2 - r$ at the point $\left(2, \dfrac{\pi}{3}\right)$ (see Fig. 3.6).

1. $q_r = \dfrac{1}{r} \dfrac{\partial \psi}{\partial \theta} = \dfrac{1}{r}\,(2r\theta) = 2\theta = \dfrac{2\pi}{3} = 2.095$ ft per sec

2. $q_\theta = -\dfrac{\partial \psi}{\partial r} = -\theta^2 + 1 = -\dfrac{\pi^2}{9} + 1 = -0.096$ ft per sec

3. $q = \sqrt{q_r^2 + q_\theta^2} = \sqrt{(2.094)^2 + (-0.096)^2} = 2.10$ ft per sec

4. $\theta' = \tan^{-1} \dfrac{q_\theta}{q_r} = \tan^{-1} 0.0479 = 2.63°$, or $57.37°$ *to the horizontal*

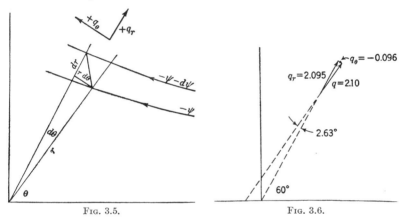

Fig. 3.5.   Fig. 3.6.

Before closing these sections on the derivations of the partial derivatives of the stream function in order to get the velocity components, it is important to note that it was not necessary to assume a velocity down and to the right in the proofs of Sects. 3.7 and 3.8. However, if other velocities and quadrants are assumed, considerable thought must be paid to sign conventions for dx and dy; the manner in which the flow enters the control zone; and the way in which the zone is constructed.

In the following pages, numerous different stream functions will be derived, and, as convenience suits us, rectilinear, polar, or mixed coordinates may be used (see Sect. 2.11). It must always be realized that any stream function expressed in one set of coordinates can also be expressed in others.

3.9. Stream Functions for Uniform Flow. The stream functions for uniform flow usually carry capital letters instead of small ones: a steady horizontal flow to the right is expressed by $\psi = Vy$; one to the left by $\psi = -Vy$. Flows may be combined to give steady flows at specified angles to the x axis. When this is done, however, actual numbers are usually employed as some particular angle is desired.

Example 3.3. Find ψ for a steady flow of 50 ft per sec down and to the right at $36°52'$ to the x axis.

For flow to the right, u is plus. For downward flow v is minus.

$$\tan \theta = \tan 323°8' = -0.750 = \frac{v}{u}$$

and hence

$$v = -0.750u$$

Also

$$u^2 + v^2 = 50^2$$

so that, solving,

$$u = 40 \text{ ft per sec}$$
$$v = -30 \text{ ft per sec}$$
$$u = \frac{\partial \psi}{\partial y} = 40; \qquad \psi_1 = 40y$$
$$v = -\frac{\partial \psi}{\partial x} = -30; \qquad \psi_2 = 30x$$
$$\psi_3 = \psi_1 + \psi_2 = 40y + 30x$$

The principle involved in adding the stream functions will be discussed later

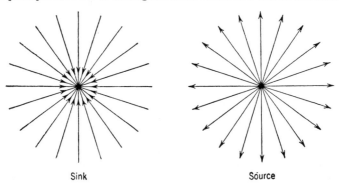

Sink   Source

Fig. 3.7.   Streamlines for a source and a sink.

3.10. The Stream Function for a Source. Of considerable use in later developments is a tiny opening from which fluid emanates radially at a uniform rate. Such a point is called a *source*. The opposite opening— one that absorbs fluid—is called a *sink*. While these devices are not found in nature, their usefulness for making possible the mathematical representation of numerous flow patterns has been amply demonstrated.

Since an infinite velocity is required for a finite amount of fluid to enter or leave a "point," sources and sinks are classed as *singular points*. Special care must be taken in applying the laws of flow to regions containing such points. Both sources and sinks are defined by their strengths, *i.e.*, the volume of fluid they emanate or absorb per second, Q.

The stream functions equations already developed can be used in developing ψ for a source. We have already defined the flow as being radial, and hence $q_\theta = 0$. If Q sq ft per sec emanate from the source, the radial velocity at radius r will be $q_r = Q/2\pi r$. Hence we have

$$q_r = \frac{1}{r}\frac{\partial \psi}{\partial \theta} = \frac{Q}{2\pi r}$$

and (setting $\psi = 0$ when $\theta = 0$)

$$\psi = \frac{Q\theta}{2\pi} \tag{3.13}$$

As a check we note

$$q_\theta = -\frac{\partial \psi}{\partial r} = 0$$

according to our definition of radial flow only.

While we shall be largely concerned with two-dimensional flow for

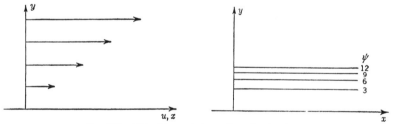

FIG. 3.8. Velocities and streamlines for $\psi = 3y^2$.

some time, it is of interest to note that the radial velocity for a three-dimensional source is developed in the same way; that is, $q_r = Q/4\pi r^2$.

3.11. Plotting by Using ψ. When ψ is known for a flow, we may use Eq. (3.11) or (3.12) to get the velocities. If, then, some particular points are selected and the velocities found for those points, a graphical representation may be made showing some velocities and giving a general idea of the flow picture.

A much more complete representation may be made by selecting values for ψ and plotting the streamlines (see Figs. 3.8 and 3.9). If even increments of ψ are assumed, the stream channels will each contain equal amounts of fluid. Then locations where the stream channels neck down are points of high velocity and, in turn, low pressure. Conversely a widening of a stream channel denotes a rising pressure.

It is entirely possible to find the expression for the flow about any reasonable airfoil and by means of Bernoulli's equation then to find the pressure and the velocity distribution.

3.12. Source in a Uniform Stream.   The stream functions of greatest use are those which represent the sum of several simple flows, and, for

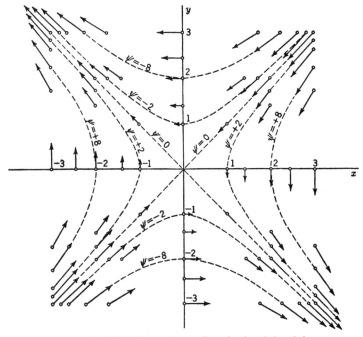

Fig. 3.9.   Velocities and streamlines for $\psi = 2x^2 - 2y^2$.

reasons that will be considered at a later time, the stream function of a complex flow is simply the sum of the stream functions of its components. The first and simplest of these combinations is that of a source $\psi_1 = Q\theta/2\pi$ and a uniform flow to the left $\psi_2 = -Vy$ being combined to yield a source in a uniform stream $\psi_3 = \psi_1 + \psi_2 = Q\theta/2\pi - Vy$. It will develop that the addition may be performed graphically or algebraically, the procedure being best shown by an example.

Example 3.4.   Plot the streamlines of a source ($Q = 2,000$ sq ft per sec) in a uniform stream ($V = 100$ ft per sec to the left).

For this case $\psi = 1,000(\theta/\pi) - 100y$. The graphical method will be outlined first.   Consider $\psi_1 = 1,000(\theta/\pi)$.   If we consider θ increments of $18°$ each, we shall have

θ, deg	$\dfrac{\theta}{\pi}$	$\psi_1 = 1,000\,\dfrac{\theta}{\pi}$
0	0	0
18	0.1	100
36	0.2	200
54	0.3	300
72	0.4	400
90	0.5	500
108	0.6	600
126	0.7	700

The lines of constant ψ_1 are shown dotted in Fig. 3.10.   Now consider

$$\psi_2 = -Vy - 100y.$$

y	$-100y = \psi_2$
0	0
1	-100
2	-200
3	-300
4	-400
5	-500
6	-600

These lines are also shown in Fig. 3.10.   To find, for instance, $\psi_3 = 0$, we need only connect the points of $\psi_1 = +100$ and $\psi_2 = -100$; $\psi_1 = +200$ and $\psi_2 = -200$, etc.   In a similar manner we may plot $\psi_3 = +100$ or -100 or, indeed, any values desired.

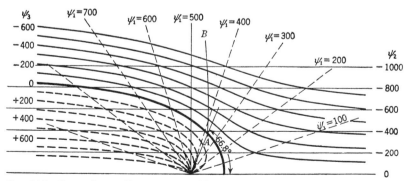

Fig. 3.10.   Source in a uniform stream.   *AB* is the locus of maximum vertical velocity points.

The second, or algebraic, method of plotting is accomplished by assuming values of ψ_3 and solving the original equation. For instance,

$$\psi_3 = 1,000\,\frac{\theta}{\pi} - 100y = 0$$

yields

$$y = 10\,\frac{\theta}{\pi}$$

From this we get

θ, deg	$\dfrac{\theta}{\pi}$	$y = \dfrac{10\theta}{\pi}$
0	0	0
18	0.1	1.0
36	0.2	2.0
54	0.3	3.0
72	0.4	4.0
90	0.5	5.0
108	0.6	6.0

It will be seen in Fig. 3.10 that this curve of $\psi_3 = 0$ is identical with the one found graphically. In general, either graphical or algebraic plotting may be performed, but in a preponderance of the instances only the latter will be practical.

Now, since no fluid crosses a streamline, any streamline may be replaced by a solid body. An interesting case is that of $\psi_3 = 0$.* This line divides the source flow from that of the uniform flow, the latter being forced up as though flowing over a bluff. Thus the source serves as a mathematical device for representing the bluff. Conversely, if it is desired to find the velocities and pressures over an object of form similar to $\psi_3 = 0$, the procedure follows this same pattern.

It is now in order to examine several of the relations that develop from the equation $\psi_3 = (Q\theta/2\pi) - Vy$.

1. *Maximum Height of $\psi = 0$.* The maximum height of the $\psi = 0$ streamline may be found by noting that under that condition $y = Q\theta/2\pi V$, and since Q and V are constant for a given case, the maximum height occurs when $\theta = \pi$. It is then equal to $Q/2V$ and is given the symbol h.

---

* At a later time when several flows are to be added, the values of the various ψ lines will be most easily understood if they are considered as representing the volume of fluid between those lines and an arbitrary reference. Thus $\psi = 0$ does not have a special significance because it is $\psi = 0$ but because in this case it is the line separating the source flow from the uniform flow. If the source had been put at $y = 2$ instead of on the x axis, the line $\psi = -200$ would be the one of greatest interest. We shall in general select our references so that $\psi = 0$ does separate important zones.

2. *Stagnation Point.* The velocity components at any point in a flow may be considered as the sum of the flow components at the point. In this case the horizontal component of the source flow is

$$u = \frac{Q}{2\pi r} \cos \theta = \frac{Q}{2\pi r} \frac{x}{r}$$

and the vertical component is

$$v = \frac{Q}{2\pi r} \sin \theta = \frac{Q}{2\pi r} \frac{y}{r}$$

The uniform flow has a horizontal component only of the amount

$$u = -V$$

Thus the total flow is

$$u = -V\left(1 - \frac{hx}{\pi r^2}\right)$$

and

$$v = \frac{Vhy}{\pi r^2}$$

The stagnation point is realized when $u = v = 0$, or, from the above equation, when $y = 0$, or, in polar coordinates, when $\theta = 0$. Recalling that $r^2 = x^2/\cos^2 \theta$, we find by setting $u = 0$ that the stagnation point occurs when $x = h/\pi$, and it will not move with changing flow conditions as long as the ratio Q/V remains constant.

3. *Lines of Constant Vertical Velocity.* Solving the expression for the vertical velocity for r, we get

$$r^2 = \frac{Vhy}{\pi v} = x^2 + y^2$$

and hence the vertical velocity v is a constant along circles whose centers are on the y axis.

4. *Line of Maximum Vertical Velocity.* For soaring, the location of the maximum vertical velocity becomes of interest.

$$v = \frac{Vhy}{\pi r^2} = \frac{Vhr \sin \theta}{\pi r^2}$$

For

$$\psi_3 = 0, \ y = \frac{Q\theta}{2\pi V} = \frac{h\theta}{\pi} = r \sin \theta$$

and hence

$$v = \frac{V \sin^2 \theta}{\theta}$$

Taking

$$\frac{dv}{d\theta} = 0$$

we get

$$2\theta = \tan\theta$$

and

$$\theta = 66.8°$$

Hence the maximum vertical velocity at the surface of the bluff occurs where a line at 66.8° to the horizontal intersects the $\psi = 0$ curve. Further values of $v_{(max)}$ are shown in Fig. 3.10 along the line AB.

5. *Point of Maximum Velocity.* Possibly the parameter of greatest importance is the maximum increase of free-stream velocity as the uniform flow passes $\psi_3 = 0$. In this case we have

$$q = \sqrt{u^2 + v^2} = V\sqrt{\left(1 - \frac{hx}{\pi r^2}\right)^2 + \left(\frac{hy}{\pi r^2}\right)^2}$$

which becomes in polars

$$\frac{q}{V} = \sqrt{\left(1 - \frac{Q}{2\pi V}\frac{r\cos\theta}{r^2}\right)^2 + \left(\frac{Q}{2\pi V}\frac{r\sin\theta}{r^2}\right)^2}$$

For $\psi_3 = 0$,

$$Q = \frac{2\pi V r \sin\theta}{\theta}$$

and hence

$$\frac{q}{V} = \sqrt{1 - \frac{\sin 2\theta}{\theta} + \frac{\sin^2\theta}{\theta^2}} \tag{3.14}$$

Equation 3.14 indicates that the velocity ratio on the $\psi = 0$ line of a source in a uniform stream is independent of the source quantity and a function only of the position of the point.

To find the maximum velocity we take the first derivative and set it equal to zero:

$$\frac{d(q/V)}{d\theta} = 0 = 2\theta\sin\theta\cos\theta - \sin^2\theta - \theta^2\cos^2\theta + \theta^2\sin^2\theta$$

Assuming values of θ from 0 to 150°, a solution is found (Fig. 3.11) at 117°. This corresponds to a velocity ratio of 1.26. The minimum value of q/V is at $\theta = 0$, the stagnation point; and free stream is again attained at $\theta = 180°$, $y = h$. From Bernoulli's equation the minimum pressure is $C_p = 1 - (q/V)^2 = -0.585$, and the same minimum pressure is developed by all sources in uniform streams, regardless of the value of

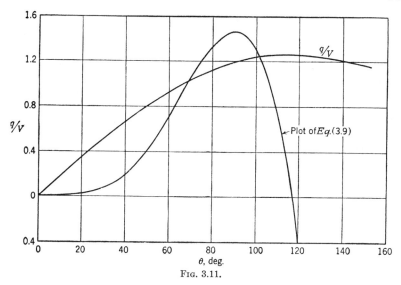

FIG. 3.11.

Q/V. The effect on the $\psi = 0$ curve of varying Q while holding V constant is shown in Fig. 3.12.

The important thing to note from the above paragraphs is that the mathematical representation of the flow has enabled us to examine a flow pattern without the need of any test apparatus. In the pages to follow this process will be repeated and enlarged.

3.13. Practical Source Applications. Many smooth curves used in aircraft approximate the $\psi_3 = 0$ curve of a source in a uniform stream. The problem of expressing the flow mathematically is then simply that of determining the proper source size and location. Molded windshields, spinners, and fuselage entries as used on twin-engined aircraft are of this

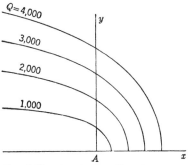

FIG. 3.12. $\psi = 0$ streamlines for several sources in a 100 ft per sec uniform stream.

type—although usually three-dimensional instead of two-dimensional. As soon as $\psi_3 = 0$ is known, the velocities and pressures at particular points of interest may be found for stress analysis.

An example is presented of a molded plastic windshield, assumed to be two-dimensional. The proper source size and location for one speed are found; and the equation of the windshield is then determined. The resulting knowledge that the maximum local velocity over the windshield will be only 378 mph at a free stream of 300 mph indicates that compressibility effects will be small. The velocity distribution (shown in Fig. 3.11) could be used to find the pressure distribution and hence the strength required.

Example 3.5. A windshield has the dimensions shown in Fig. 3.13 and a shape that is approximately that developed by a source in a uniform stream. Find (for a free-stream velocity of 300 mph)
 1. The equation of the windshield
 2. The point of maximum velocity
 3. The maximum velocity

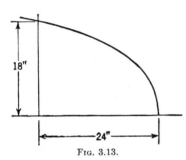

FIG. 3.13.

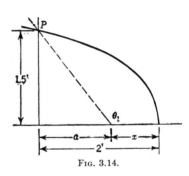

FIG. 3.14.

The unknowns are the source strength and location. Put a source of strength Q at a ft from P. Then

$$a + x = 2.0 = a + \frac{Q}{2\pi V} \tag{3.15}$$

For $\psi = 0$,

$$y = \frac{Q\theta}{2\pi V} \tag{3.16}$$

and from the geometry of the problem

$$\theta_1 = \tan^{-1}\left(\frac{-1.5}{a}\right) \tag{3.17}$$

Combining Eqs. (3.15) and (3.16),

$$a + \frac{2\pi V y}{\theta}\frac{1}{2\pi V} = 2.0$$

and canceling and inserting θ_1 from Eq. (3.17),

$$a + \frac{3}{2\tan^{-1}(-1.5/a)} = 2.0 \tag{3.18}$$

Rewrite Eq. (3.18) as

$$\tan^{-1}\left(\frac{-1.5}{a}\right) + \frac{3}{2a - 4} = 0$$

and, assuming values of a between 1.0 and 1.5, find a solution occurs when $a = 1.35$.

From Eq. (3.15), inserting a and V,

$$Q = (2.0 - a)2\pi V = 1,800 \text{ sq ft per sec}$$

and from Eq. (3.17),

$$\theta_1 = 131°58'$$

The equation of the windshield and its flow is hence

$$\psi_3 = -441y + \frac{1,800\theta}{2\pi}$$

The point of maximum velocity is 117° from the free stream, and the maximum velocity is $1.26 \times 300 = 378$ mph.

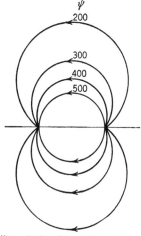

For two-dimensional curves *not* similar to the source in a uniform stream an entirely different approach is required. This other approach is discussed in Chap. 8.

3.14. Source and Sink in a Uniform Stream; the Rankine Oval. The source in a uniform flow led to an infinite boundary, but the addition of one or more sinks of total capacity equal to the source flow closes the system, and $\psi = 0$ becomes finite. The simplest of the closed systems is a single source with a single sink of equal capacity, which yields a flow as shown in Fig. 3.15 or, when combined with a uniform flow, yields patterns as shown in Fig. 3.17. The stream-line $\psi = 0$ for such flows is called a *Rankine oval*.

Fig. 3.15. Streamlines for a source and a sink, equal quantities.

Consider Fig. 3.16 wherein source A_1 is at an equal distance s from the y axis as sink A_2. The value of ψ_3 at point $P(x, y)$ for the source and sink will be

$$\psi_3 = \psi_1 + \psi_2 = \frac{Q\theta_1}{2\pi} - \frac{Q\theta_2}{2\pi}$$

and from Fig. 3.16

$$\psi_3 = \frac{Q}{2\pi}(\theta_1 - \theta_2) = \frac{Q\phi}{2\pi}$$

Thus, for $\psi_3 = \text{const}$, ϕ is a constant, and the streamlines ψ_3 become circles with their centers on the y axis. Now

$$\tan \theta_1 = \frac{y}{x - s} \qquad \text{and} \qquad \tan \theta_2 = \frac{y}{x + s}$$

and since

$$\tan \phi = \tan (\theta_1 - \theta_2) = \frac{\tan \theta_1 - \tan \theta_2}{1 + \tan \theta_1 \tan \theta_2}$$

we have

$$\tan \phi = \frac{2ys}{x^2 + y^2 - s^2}$$

and

$$\psi_3 = \frac{Q}{2\pi} \tan^{-1} \frac{2ys}{x^2 + y^2 - s^2}$$

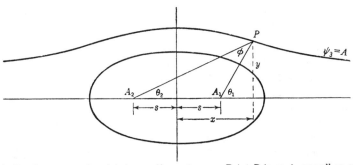

FIG. 3.16.   A source and a sink in a uniform stream.   Point P is on ψ_3 as well as on ψ_4.

Adding the uniform flow, we get

$$\psi_4 = -Vy + \frac{Q\phi}{2\pi} \tag{3.19}$$

Equation 3.19 is the stream function of a source and a sink in a uniform stream, the distance between source and sink being $2s$. $\psi_4 = 0$ is the previously mentioned oval.

The boundary $\psi_4 = 0$ becomes

$$y = \frac{Q}{2\pi V} \tan^{-1} \frac{2ys}{x^2 + y^2 - s^2} = \frac{Q}{2\pi V} \phi$$

and

$$\phi = \frac{2\pi V b}{Q}$$

When $x = 0$, $y = b$ (where b is the height of the semi-minor axis), and we then have

$$\frac{b}{s} = \cot \frac{\phi}{2} = \cot \frac{\pi V b}{Q} \tag{3.20}$$

which may be used to determine the height of the semi-minor axis once the other parameters have been determined.

We may determine the length of the semi-major axis from the fact that $(a, 0)$ is a stagnation point. At $(a, 0)$ the uniform flow is equal and opposite to the horizontal component of the sum of the source and the

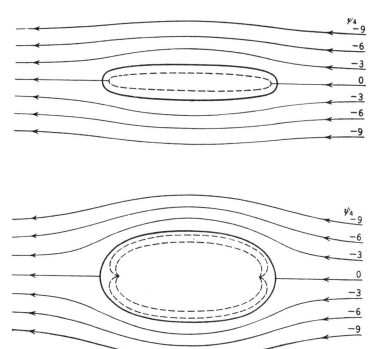

FIG. 3.17. Effect of increasing Q while holding semi-major axis constant. *Top*, $Q = 7.15$; *bottom*, $Q = 26.1$.

sink. From Fig. 3.16 and the fact that $q_r = Q/2\pi r$ for a source and $-Q/2\pi r$ for a sink we have

$$u = 0 = -V + \frac{Q}{2\pi}\left(\frac{1}{a - s} - \frac{1}{a + s}\right)$$

or

$$V = \frac{Qs}{\pi(a^2 - s^2)}$$

from which

$$\frac{a^2}{s^2} = 1 + \frac{Q}{\pi s V} \tag{3.21}$$

Thus if it is desired to represent a Rankine oval of known semi-major and semi-minor axes in a known uniform flow, Eqs. (3.20) and (3.21) may be solved for the proper source size and location.

The behavior of the flow pattern if the semi-major axis and uniform flow are held constant and the source strength varied is shown in Fig. 3.17. The ovals of $\psi_4 = 0$ are not ellipses, however, being somewhat blunter. This is shown in Fig. 3.18.

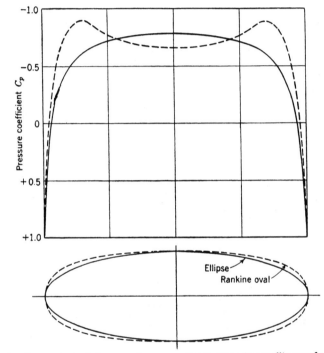

FIG. 3.18. Comparison of shapes and pressure distributions for an ellipse and a Rankine oval, both with a 3:1 fineness ratio.

Later on (in Sect. 6.2) a discussion of desirable pressure distribution is made, and it is there pointed out that low-pressure peaks near the leading edge are undesirable from both a boundary-layer stand point and critical-speed criteria. With these facts in mind, a comparison of pressures over a Rankine oval and over an ellipse takes on real significance. Figure 3.18 demonstrates how the bluntness of the oval causes regions of excessively low pressure distribution near the leading and trailing edges, while the ellipse is near perfect for an even flow. Ellipses are hence used for airfoil leading edges and are superior aerodynamically to Rankine ovals. It is of interest to note that Rankine himself felt that his ovals, being "mathematical" shapes, were specially advantageous.

Of more interest than the oval for future work is the shape obtained for the $\psi = 0$ line when the spacing between an equal quantity source and sink is decreased while the product of their strength and the distance between them is held constant. From Fig. 3.19 it is seen that, as $s \to 0$, $a/b \to 1$; that is, the oval approaches a circle as the source approaches the sink.

The Rankine oval does not have a single point of maximum velocity for all ratios as did the source in a uniform flow, being then at 117° from the oncoming free stream. Instead, two points of maximum velocity (q being equal for both) exist; and these move from the 117° point toward the 90° mark as the sink approaches the source, reaching the 90° mark to form one maximum-velocity point when the distance between source and sink is infinitesimally small. As this procedure is taking place the maximum velocity is rising until it reaches $2V$ at the

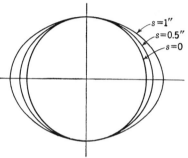

Fig. 3.19. $\psi = 0$ streamlines, showing the effect of decreasing source and sink spacing s while holding semi-minor axis constant.

limit, and C_p then becomes -3.0. This is to be discussed in the next section.

Example 3.6. Find the source and sink strength, and their spacing, and the equation of a Rankine oval having a major axis of 8 in. and minor axis of 3 in. $V = 500$ ft per sec.

Solving Eq. (3.21) for Q, we have

$$Q = \frac{a^2 - s^2}{s} \pi V \qquad (3.22)$$

and, substituting into Eq. 3.20, this yields

$$\frac{b}{s} = \cot \frac{\pi V b}{Q} = \cot \frac{bs}{a^2 - s^2} \qquad (3.23)$$

Writing

$$y_1 = \frac{b}{s} - \cot \frac{bs}{a^2 - s^2}$$

and inserting the values for the semi-major and semi-minor axes in feet we get

$$y_1 = \frac{0.125}{s} - \cot \frac{0.125s}{(0.33^2 - s^2)}$$

and substituting reasonable values of s and plotting, we find $y_1 = 0$ when $s = 0.284$ ft.

Putting s, a, and V into Eq. (3.22) yields $Q = 170$ sq ft per sec. The equation of the oval becomes

$$\psi_4 = -500y + \frac{170}{2\pi} \tan^{-1} \frac{0.568y}{x^2 + y^2 - 0.0808}$$

3.15. Doublet in a Uniform Stream: the Circular Cylinder. Consider the case when the source and the sink approach each other while the product of the source strength Q and distance between the source and sink $2s$ remains constant. Write the "doublet" strength equal to this product; that is, $\mu = 2Qs$.

As s becomes very small, the stream function for a source and a sink then tends to the limit as $s \to 0$ of

$$\psi = \frac{2s}{2s} \frac{Q}{2\pi} \tan^{-1} \frac{2ys}{x^2 + y^2 - s^2} = \frac{\mu}{4\pi s} \frac{2ys}{x^2 + y^2}$$

$$\psi = \frac{\mu}{2\pi} \frac{y}{x^2 + y^2} = \frac{\mu}{2\pi} \frac{\sin \theta}{r} \qquad (3.24)$$

A source and sink whose spacing tends to zero while the product $2Qs$ remains finite is called a *doublet of strength μ*, and the very short line joining the source and sink is called its *axis*.

The streamlines due to a doublet at the origin are circles that pass through the doublet and are tangent to its axis. A plot of a doublet is shown in Fig. 3.20.

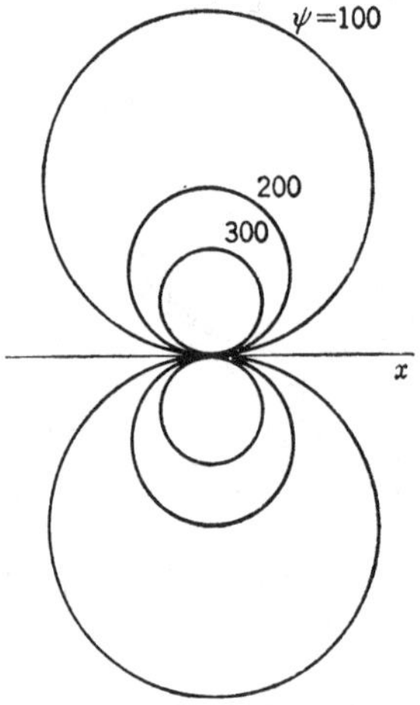

FIG. 3.20. Streamlines for a doublet.

Adding the stream function for a uniform flow to that of a doublet yields

$$\psi = -Vy + \frac{\mu}{2\pi} \frac{y}{x^2 + y^2} \qquad (3.25)$$

For the case of $\psi = 0$, $x^2 + y^2 = \mu/2\pi V$, which, since both μ and V are constant for a given case, represents a circle at the origin of radius $a = \sqrt{\mu/2\pi V}$ or $\mu = 2\pi V a^2$. Inserting this value in Eq. (3.25), and using polar representation, we have

$$\psi = -V \sin \theta \left(r - \frac{a^2}{r} \right) \qquad (3.26)$$

for the flow past a circular cylinder of radius a. A plot of Eq. (3.26) is shown in Fig. 3.21.

The local velocities may now be found from Eq. (3.12).

$$q_r = \frac{1}{r}\frac{\partial \psi}{\partial \theta} = -V \cos \theta \left(1 - \frac{a^2}{r^2}\right) \tag{3.27}$$

$$q_\theta = -\frac{\partial \psi}{\partial r} = V \sin \theta \left(1 + \frac{a^2}{r^2}\right) \tag{3.28}$$

At the surface of the cylinder $q_r = 0$ and $r = a$, yielding

$$q_\theta = 2V \sin \theta \tag{3.29}$$

Obviously q_θ is a maximum at 90° when it has the value $2V$. That is, fluid passing over a circular cylinder must theoretically accelerate to

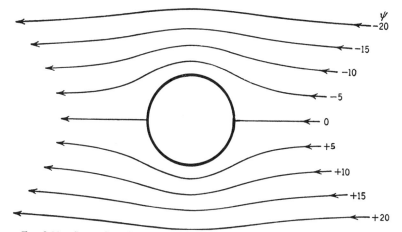

FIG. 3.21. Streamlines for a circular cylinder in a uniform stream, no circulation.

twice the free-stream velocity. In practice, the viscous effects are so pronounced that a value somewhat below 2.0 is realized. The exact value attained varies with Reynolds number, initial turbulence in the air, the rigidity of the circular-cylinder mounting, and other items.

As a matter of additional interest, the speed at which the speed of sound is first attained locally is called the *critical Mach number* and would first be reached on a circular cylinder at $V_0 = \frac{1}{2}V_c$, where V_c is the speed of sound if no compressibility existed.* For a perfect compressible fluid (that is, inviscid) the critical Mach number becomes 0.409 for a circular cylinder and 0.573 for a sphere.

To return to the problem at hand, the local pressures may be found

* This supposition is, of course, farfetched, as the existence of a sound wave is predicated on the existence of compressibility of the fluid.

directly by writing Bernoulli's equation between the free stream V and a point on the cylinder where only q_θ exists. Thus, from Eq. (3.10),

$$C_p = 1 - \left(\frac{q_\theta}{V}\right)^2 \qquad (3.30)$$

Substituting for q_θ, using Eq. (3.29), we get

$$C_p = 1 - 4 \sin^2 \theta \qquad (3.31)$$

from which we see that the minimum pressure coefficient is -3.0, which occurs at $\pi/2$. The pressure distribution is, by Eq. (3.31), symmetrical about $x = 0$ and $y = 0$, and hence there is no resultant force on the cylinder. This is, of course, at variance with experiment wherein a large drag appears, and (as will be seen) a large lift may also be evident. This great failure of the theory was first advanced by d'Alembert and is known as *d'Alembert's paradox*. The solution was advanced by Helmholtz (1821–1894) in 1858, nearly a hundred years later. It is discussed in the next chapter.

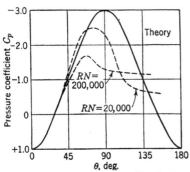

FIG. 3.22. Static pressure distribution over a circular cylinder, theory and experiment.

To continue with the present case, the experimental values of the pressure as well as the theoretical values are shown in Fig. 3.22, where it is seen that the agreement between the two is good over the upstream 40°. It will also be seen that the experimental pressure over the downstream 80° increases with increasing Reynolds number so that the drag coefficient decreases in that range. A similar phenomenon is observed with spheres where near RN $= 420,000$ in free air the pressure pattern changes so rapidly that the drag as well as the drag coefficient decreases. This extreme function of Reynolds number is used in measuring wind tunnel turbulence.

An indirect use of the doublet in a uniform flow, considering the static pressures ahead of a circular cylinder, yields the position error of a pitot-static tube placed in front of a round mounting strut. This arises as follows:

It is obvious that no vertical velocity exists along the x axis, and for that horizontal line Eq. (3.27) reduces to

$$q_r = -V\left(1 - \frac{a^2}{r^2}\right) \qquad (3.32)$$

From Eq. (3.31) the local pressure is then

$$C_p = 1 - \left(1 - \frac{a^2}{r^2}\right)^2 \qquad (3.33)$$

It is seen immediately that this expression quite properly goes to zero at $r = \infty$ and to $+1.0$ at $r = a$.

The accuracy of Eq. (3.33) agrees well with practice, and similar relations empirically derived are used to design pitot-static tubes so that the static orifices are properly adjusted for the effect of the stem behind them. The effect of a circular stem 1.7 diameters behind the static orifices is seen to be $C_p = +16.5$ per cent from Eq. (3.33). In practice Spaulding and Merriam (Ref. 3.4) found an 18 per cent error.

PROBLEMS

3.1. Is the flow represented by $\psi = -3x^2y + 3xy^2$ incompressible?

3.2. Is the flow represented by $\psi = -y + (y/x^2)$ incompressible?

3.3. Give the stream function for a steady flow down and to the left, 30° below the horizontal. Its velocity is 100 ft per sec.

3.4. Find the velocity components, total velocity, and flow direction at the point $(2, -4)$ for the flow represented by $\psi = 2xy + y^2$.

3.5. If $\psi = x^2y + y^2x$ represents a flow, find the velocity components, total velocity, and direction of flow at the point $(2, 3)$.

3.6. If $\psi = r\theta^2 + r^2\theta$ represents a flow, find the velocity components, total velocity, and direction of flow at the point $\left(\frac{\pi}{2}, 4\right)$.

3.7. Sketch the streamlines in all quadrants for $\psi = x^2y$.

3.8. If $\psi = r\theta^2$, find the velocity components and total velocity at the point $\left(\frac{\pi}{2}, 3\right)$.

3.9. If a source in a uniform stream has a maximum height of 10 ft for the $\psi = 0$ lines, how far is the stagnation point from the source?

3.10. If a source is in a uniform stream of 100 ft per sec and the sum of the maximum height and the distance from the source to the front stagnation point is 50 ft, what must the source quantity be?

3.11. For the conditions of Prob. 3.10 state the maximum velocity to be found anywhere in the field and the point where it will be found. What will the pressure coefficient be at that point?

3.12. A circular cylinder is in a 100-mph stream at standard conditions. Calculate the pressure coefficients that would theoretically be found at points 0, 30 and 90° from the flow direction.

3.13. If the cylinder of Prob. 3.12 is 1 in. in diameter, find the pressure coefficient for a point 3 in. ahead of it.

3.14. At what point on a circular cylinder in a uniform stream is the tangential velocity one-half the free-stream value?

REFERENCES

3.1. W. F. Durand, "Aerodynamic Theory," Vol. I, pp. 306–394, Verlag Julius Springer, Berlin, 1934.

3.2. H. Glauert, "Aerofoil and Airscrew Theory," pp. 18–31, Cambridge, University Press, London, 1926.

3.3. V. L. Streeter, "Fluid Dynamics," Chap. VI, McGraw-Hill Book Company, Inc., New York, 1948.

3.4. Kenneth G. Merriam and Ellis R. Spaulding, Comparative Tests of Pitot-static Tubes, *TN* 546, 1935, p. 21.

CHAPTER 4

OTHER FLOW FUNCTIONS

4.1. Vortex Motion. The existence of a fluid motion in which the fluid whirls about some axis is well known in nature. Waterspouts, whirlpools, and tornadoes are convenient examples. But the connection between these phenomena and the lift generated dynamically by a body in a fluid is not at all obvious, and it is not surprising that perhaps a century had to pass between the sound foundation of the perfect-fluid theory and the addition to it that explained lift. This contribution was

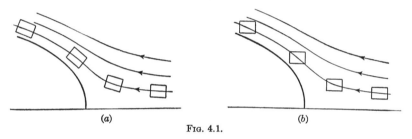

| (a) | (b) |

FIG. 4.1.

summed up by Helmholtz in 1858, who formulated *vortex laws of motion*. Actually, these dealt with the three-dimensional vortex and are discussed in Chap. 9. A two-dimensional vortex will be developed in the following paragraphs.

The first physical concept to grasp is the manner in which a fluid stream negotiates a curved path. The particles* do not "line up" with the path (Fig. 4.1a) because in a perfect fluid there are no shear forces to make them rotate. Instead they maintain their orientation (Fig. 4.1b) in a manner similar to that of a car on a Ferris wheel with, however, a little shape distortion not shown in Fig. 4.1.

The significance as to whether the particles "rotate" or not has a deeper underlying value not readily apparent. In our derivation of Bernoulli's equation no restriction was placed on the type of flow in the stream tube considered, and it follows that Bernoulli's constant holds anywhere along a given streamline, whether the motion be *rotational* or

* A fluid particle is another name for a small reference zone which moves with the fluid, and, although its shape may change, contains the same amount of fluid over a period of time.

irrotational. However, it is shown in more advanced work that only if the motion is irrotational will Bernoulli's constant hold throughout the entire field.

In practice, irrotational motion is found as long as boundary layers, turbulence, and (in the case of a compressible fluid) shock waves are avoided. It will be seen in the following pages that a good definition of irrotational motion is "motion such that $\nabla \times \mathbf{q} = 0$ at every point."

4.2. Circulation. A second new term associated with vortex motion is *circulation*, defined as the line integral of the tangential velocity taken around a closed curve. It is usually called Γ, square feet per second.

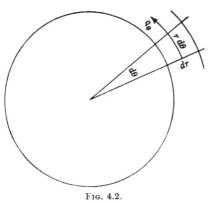

Circulatory (irrotational) motion exists outside of the rotational core of a vortex in a real fluid. The particles in circulatory motion do not rotate, even when they follow a circular path. Let us consider this case in more detail. Assume fluid with pure circulation (tangential velocity only) moving in a circular path as shown in Fig. 4.2.

The fluid in a zone $r\, d\theta\, dr$ will suffer a centrifugal force in the amount $(m/r)\, q_\theta^2$ that must be balanced by an increase in pressure force that acts in a centripetal (inward) direction. Thus

FIG. 4.2.

$$F = \text{pressure} \times \text{area}$$

$$\rho\, dr\, r\, d\theta\, \frac{q_\theta^2}{r} = dp\, r\, d\theta$$

or

$$\frac{dp}{dr} = \rho\, \frac{q_\theta^2}{r} \tag{4.1}$$

Now we have stipulated that the particles must not rotate in order that the head remain constant and Bernoulli's equation be applicable. We therefore write Eq. (3.8) for a velocity q_θ and differentiate it with respect to r.

$$\frac{dp}{dr} + \rho q_\theta \frac{dq_\theta}{dr} = 0$$

Substitution from Eq. (4.1) yields

$$\rho\, \frac{q_\theta^2}{r} + \rho q_\theta \frac{dq_\theta}{dr} = 0$$

which integrates to

$$rq_\theta = \text{const} \tag{4.2}$$

Thus the velocity in pure circulatory irrotational flow varies inversely as the radius. Equation (4.2) tells us in addition that when the radius is zero the velocity would be infinite. Nature avoids the infinite velocity by having a central vortex core rotate like a wheel—with aerodynamic rotation of the particles. In the case of the vortex in air the velocity follows Eq. (4.2) beautifully until a certain radius r_1 is reached. (The size of r_1 is discussed in Chap. 9.) Inside $r = r_1$ the core revolves, making the complete velocity distribution as shown in Fig. 4.3. Outside of r_1 the circulation remains constant [Eq. (4.2)], and the motion is irrotational. At r_1 both the

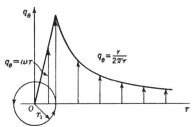

Fig. 4.3. Distribution of tangential velocity for a practical vortex.

rotating core and the circulatory flow have equal velocities so that we may write

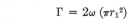

$$\frac{\Gamma}{2\pi r_1} = \omega r_1$$

where ω = rotational speed of the core, radians per sec.

Solving for the circulation Γ, we have

$$\Gamma = 2\omega\,(\pi r_1{}^2)$$

Thus the circulation may be thought of as consisting of area $\pi r_1{}^2$ rotating at an angular velocity (called the *vorticity*) of 2ω. In a perfect fluid the core area may be considered as exceedingly small, and the vorticity exceedingly large to yield any specified circulation. Another way to discuss circulation is to refer to it as the *vortex strength*, and it follows that the circulation about a given closed path is equal to the sum of the vortex

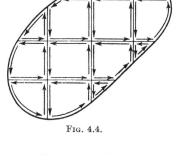

Fig. 4.4.

strengths contained within the path (Stokes' theorem). This conclusion is simply demonstrated by dividing any arbitrary area containing vortices into a grid (Fig. 4.4) and noting that the internal components of circulation cancel out.

From the above discussion it is seen that, if the circulation is zero about a region containing fluid only, there is no net rotation in the region.

We may now proceed to derive the stream function for circulatory flow. Vortex motion is, intuitively, motion without a radial component. The only velocity that exists is q_θ, and the distance through which it exists is $2\pi r$. Hence, from the definition of circulation,

$$\Gamma = 2\pi r q_\theta$$

or

$$q_\theta = \frac{\Gamma}{2\pi r} \tag{4.3}$$

But we have already derived an expression for q_θ in terms of the stream function which is

$$q_\theta = -\frac{\partial \psi}{\partial r} \tag{4.4}$$

and hence, equating and integrating,

$$\psi = -\frac{\Gamma}{2\pi} \ln r + C \tag{4.5}$$

where C = constant of integration.

The value of C may be determined as follows: In the next section we are going to add circulation to the flow about a circular cylinder of radius a, and it becomes most convenient to let the $\psi = 0$ curves of both flows coincide. Thus, if in Eq. (4.5) we set $\psi = 0$ when $r = a$, we get, finally,

$$\psi = -\frac{\Gamma}{2\pi} \ln \frac{r}{a} \tag{4.6}$$

There is no restriction as to the size of a, which can be made infinitesimally small or exceedingly large. Nor is a the vortex core r_1 previously mentioned as a phenomenon that occurs in a real fluid. The term a in this case is best considered as an arbitrary constant, selected for convenience. If we fail to use the same a for both the vortex and the circular cylinder when we add the two flows, the resulting flow patterns are perfectly acceptable but the streamline that represents the cylinder is no longer $\psi = 0$. Indeed, even though identical values are used for cylinder and circulation, it will later be seen that "excessive" circulation moves the $\psi = 0$ streamline off the cylinder. The flow inside the cylinder may be examined by a method to be shortly presented and found to be irrotational at all points except $r = 0$.

Many writers use a *unit vortex* ($a = 1.0$) so that the stream function becomes

$$\psi_{\text{unit vortex}} = -\frac{\Gamma}{2\pi}\ln r$$

It is also convenient to note that any closed curve may be warped into a circle by replacing short segments of the curve by circular arcs and radial segments. From Eq. (4.5) there is no radial velocity due to a vortex; hence the radial segments have no circulation, and the entire value remains in the arcs. For more complex types the circulation may be written (see Fig. 4.5)

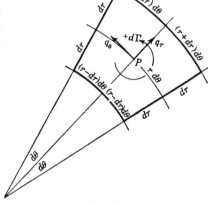

Fig. 4.5.   Fig. 4.6.

$$\Gamma = \oint q \cos \theta \, ds \tag{4.7}$$

or, in vector notation,

$$\Gamma = \oint \mathbf{q} \cdot d\mathbf{s} \tag{4.8}$$

Now it is convenient to have a rapid method of determining whether or not a flow is irrotational so that we may know whether or not Bernoulli's equation is applicable and for a second important reason to be discussed later. Such a "checking" relation may be developed by summing the circulation due to any vorticity present about an arbitrary zone and examining the conditions which must arise if the vorticity is to be zero.

Thus, using polar coordinates for a flow with tangential velocity q_θ and radial velocity q_r, and a zone such as shown in Fig. 4.6, we have

$$d\Gamma = \left(q_\theta + \frac{\partial q_\theta}{\partial r}\, dr\right)(r + dr)\, 2d\theta + \left(q_r - \frac{\partial q_r}{\partial \theta}\, d\theta\right) 2dr$$
$$- \left(q_\theta - \frac{\partial q_\theta}{\partial r}\, dr\right)(r - dr)\, 2d\theta - \left(q_r + \frac{\partial q_r}{\partial \theta}\, d\theta\right) 2dr$$

Expanding and clearing,

$$dΓ = 4r \, dθ \, dr \left(\frac{1}{r} q_θ + \frac{\partial q_θ}{\partial r} - \frac{1}{r} \frac{\partial q_r}{\partial θ} \right)$$

Dividing through by the area to get the vorticity, we have

$$\text{Vorticity} = \frac{1}{r} q_θ + \frac{\partial q_θ}{\partial r} - \frac{1}{r} \frac{\partial q_r}{\partial θ} \tag{4.9}$$

If the circulation is zero, Eq. (4.9) becomes

$$\frac{1}{r} q_θ + \frac{\partial q_θ}{\partial r} - \frac{1}{r} \frac{\partial q_r}{\partial θ} = 0 \tag{4.10}$$

From Eq. (3.12) this becomes

$$\frac{1}{r} \frac{\partial ψ}{\partial r} + \frac{\partial^2 ψ}{\partial r^2} + \frac{1}{r^2} \frac{\partial^2 ψ}{\partial θ^2} = 0 \tag{4.11}$$

which may be used to determine whether the flow represented by a certain $ψ$ is irrotational or not—with one exception. It does not furnish a check for the theoretical condition of infinite vorticity over an infinitesimally small point at $r = 0$, and, it so develops, this particular case is important. It will be considered in detail later. To recapitulate, if Eq. (4.11) is satisfied by a particular relation for $ψ$, then at least the flow outside of $r = 0$ is irrotational and in addition will be noncirculatory unless a point vortex exists at $r = 0$.

The solution to the problem of treating the rotation of the vortex core is very simple. If a solid body is used to replace the core, then the *fluid* can have irrotational motion everywhere. Thus a flow with circulation can be *completely* irrotational as long as the circulation occurs about a solid body. As far as Eq. (4.10) is concerned, it will remain zero because there is no rotation in the fluid. That is, when a body is present, Eqs. (4.10) and (4.11) are checks for zero rotation but not zero circulation. When no body is present, they are satisfied only by zero rotation and zero circulation if no point vortex exists at $r = 0$. If one does, then as previously mentioned there will be circulation around the point $r = 0$, which will itself be a singular point.

Equations (4.10) and (4.11) are applicable to pure circulatory motion as well as more complex forms. In compressible flow of the type that is constant along a radius Eq. (4.10) reduces to

$$q_θ = \frac{\partial q_r}{\partial θ}$$

and has a widespread application.

The condition for irrotational flow in rectangular coordinates is as follows:

Consider a small zone in a flow of varying horizontal and vertical velocities u and v, respectively (Fig. 4.7). The circulation about the zone will be (Γ is $+$ in the counterclockwise direction)

$$d\Gamma = \left(u - \frac{\partial u}{\partial y}\,dy\right)2dx + \left(v + \frac{\partial v}{\partial x}\,dx\right)2dy - \left(u + \frac{\partial u}{\partial y}\,dy\right)2dx$$
$$- \left(v - \frac{\partial v}{\partial x}\,dx\right)2dy$$

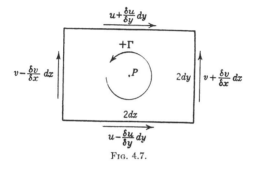

Fig. 4.7.

If $d\Gamma = 0$ (this implies that no particles with rotation are within the zone), we have

$$-\frac{\partial u}{\partial y} + \frac{\partial v}{\partial x} = 0 \qquad (4.12)$$

From Chap. 1 we may write Eq. (4.12) as

$$\nabla \times \mathbf{q} = 0 \qquad (4.13)$$

where $\mathbf{q} = iu + jv$
or, using Eq. (3.11),

$$\frac{\partial^2 \psi}{\partial x^2} + \frac{\partial^2 \psi}{\partial y^2} = 0 \qquad (4.14)$$

We may hence examine a flow for irrotation through the use of any of several equations, the particular relation selected being dependent on the form of the equation to be investigated. Equation (4.14) is called the *Laplacian* of ψ.

Now so far we have been content to "assume" a circulatory flow, and it is in order to consider how we can actually make one in the two-dimensional case. In this instance the obvious is a good answer: a circular cylinder rotated in an airstream (with sufficient end plating so that tip flow does not arise) will develop a circulation closely approximated by Eq. (4.3) in the range $q_\theta/V = 0.5$ to 3.0, an actual case being discussed in Sect. 4.5. A lifting wing will do the same, but the mechanism is a little different. It can be roughly discussed as follows:

Consider a cambered wing at zero angle of attack in a flow that has just started (Fig. 4.8a). Two particles A and B travel along their respective surfaces at *equal* speeds, and since the upper surface is "longer," A arrives at the trailing edge ahead of B. It then attempts to go around the trailing edge (b), but the general motion of the fluid sweeps it and other particles off the airfoil and B becomes a part of a *starting vortex*,

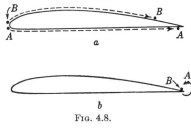

FIG. 4.8.

traveling farther and farther downstream until its effect is no longer felt at the wing. However, before it is swept downstream, it induces an increase of velocity on the airfoil upper surface that closes the gap between particles A and B and leaves a circulation on the wing equal but opposite to that of the starting vortex. This process is shown beautifully by Goldstein in Plate (9) of Ref. 4.5.

We can align this action with our assumption of very small viscosity by noting that shear forces in fluids are proportional to the velocity of adjacent layers, and hence every time the theory indicates infinite velocity the amount of shear force that then arises in our fluid of very small viscosity is enough to cause the creation of a vortex. From our mathematical treatment to follow, it will be seen that the above condition is one of infinite velocity at the trailing edge and hence amenable to the process described above.

4.3. Circular Cylinder with Circulation in a Uniform Stream. It has been shown that a doublet in a uniform stream provides a mathematical representation of a circular cylinder. Further, the pressure distribution about the circular cylinder in a perfect fluid ($C_p = 1 - 4 \sin^2 \theta$) was shown to be symmetrical about both x and y axes so that the cylinder experienced no resultant force. Let us determine the effect of adding circulation to the circular cylinder. From Eqs. (3.26) and (4.5) we have

$$\psi = -V\left(r - \frac{a^2}{r}\right)\sin\theta - \frac{\Gamma}{2\pi}\ln\frac{r}{a} \tag{4.15}$$

The velocity components are

$$q_r = \frac{1}{r}\frac{\partial\psi}{\partial\theta} = \frac{1}{r}\left[-V\cos\theta\left(r - \frac{a^2}{r}\right)\right]$$

and

$$q_\theta = -\frac{\partial \psi}{\partial r} = V \sin \theta \left(1 + \frac{a^2}{r^2}\right) + \frac{\Gamma}{2\pi r} \qquad (4.16)$$

At a point on the surface of the cylinder $r = a$, $q_r = 0$, and

$$q_\theta = 2V \sin \theta + \frac{\Gamma}{2\pi a} \qquad (4.17)$$

The symmetric field is distorted by small circulation into a pattern as shown in Fig. 4.9a, no longer symmetrical about the x axis, but still symmetrical about the y axis; that is, zero drag exists. It is seen that the zero-velocity stagnation points, formerly on the x axis, have been moved down by the circulation. This tendency continues until in the limit they meet at 270° (Fig. 4.9b). The circulation required for the stagnation points at 270° is found from Eq. (4.17):

$$q_\theta = 0 = 2V(-1) + \frac{\Gamma}{2\pi a}$$

or

$$\Gamma = 4\pi V a$$

For circulations above $\Gamma = 4\pi V a$ the stagnation point leaves the surface of the cylinder and moves vertically downward, part of the fluid circulating with the cylinder as in Fig. 4.9c.

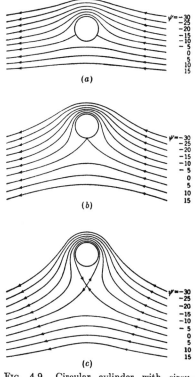

(a)

(b)

(c)

Fig. 4.9. Circular cylinder with circulation in a uniform stream of 10 ft per sec, $\Gamma = 2\pi V a$, $4\pi V a$, and $8\pi V a$. $\Gamma = 2\pi V a$ corresponds to an airfoil with $c_l = 3.14$, a very large value.

4.4. Pressures over a Cylinder with Circulation. The most important conclusion so far develops when the pressure distribution over the circular cylinder is integrated to find the resultant force. From Bernoulli's equation,

$$p = H - \tfrac{1}{2}\rho(q_\theta^2 + q_r^2) \qquad (4.18)$$

which in our case becomes ($q_r = 0$ over the cylinder)

$$p = H - \frac{1}{2}\rho \left(2V \sin \theta + \frac{\Gamma}{2\pi a}\right)^2$$

or

$$p = H - 2\rho V^2 \sin^2 \theta - \frac{\rho V \Gamma}{\pi a} \sin \theta - \frac{\rho \Gamma^2}{8\pi^2 a^2} \qquad (4.19)$$

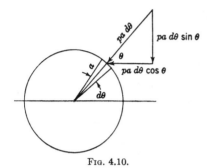

The horizontal and vertical components of the pressure force are (see Fig. 4.10)

$$X = -\int_0^{2\pi} pa \, d\theta \cos \theta \qquad (4.20)$$

and

$$Y = -\int_0^{2\pi} pa \, d\theta \sin \theta \qquad (4.21)$$

FIG. 4.10.

Substituting Eq. (4.19) in Eq. (4.20), we get

$$-X = Ha \int_0^{2\pi} \cos \theta \, d\theta - 2\rho a V^2 \int_0^{2\pi} \sin^2 \theta \cos \theta \, d\theta$$

$$- \frac{\rho V \Gamma}{\pi} \int_0^{2\pi} \sin \theta \cos \theta \, d\theta - \frac{\rho \Gamma^2}{8\pi^2 a} \int_0^{2\pi} \cos \theta \, d\theta \qquad (4.22)$$

The first and last terms integrate to $\sin \theta$ and vanish when the limits are inserted. The second term integrates to $\sin^3 \theta$, which also vanishes for the above limits; and the third term integrates to $\sin^2 \theta$ and is zero too.

$$\therefore X = 0$$

or no drag is realized.

Substituting Eq. (4.19) into (4.21), we get

$$-Y = Ha \int_0^{2\pi} \sin \theta \, d\theta - 2\rho a V^2 \int_0^{2\pi} \sin^3 \theta \, d\theta - \rho \frac{V \Gamma}{\pi} \int_0^{2\pi} \sin^2 \theta \, d\theta$$

$$- \frac{\rho \Gamma^2}{8\pi^2 a} \int_0^{2\pi} \sin \theta \, d\theta$$

The first and last terms integrate to $\cos \theta$ and vanish when the limits 0 and 2π are inserted. The second term may be written ($\sin \theta - \cos^2 \theta \sin \theta$), and it, too, integrates to cosine terms which vanish when the limits are inserted. The remainder is

$$Y = \frac{\rho V \Gamma}{\pi} \int_0^{2\pi} \sin^2 \theta \, d\theta \qquad (4.23)$$

From the double-angle trigonometric relation Eq. (4.23) becomes

$$Y = \frac{\rho V \Gamma}{\pi} \int_0^{2\pi} \left(\frac{1}{2} - \frac{\cos 2\theta}{2} \right) d\theta$$

and

$$Y = \rho V \Gamma \qquad \text{lb per ft of span} \qquad (4.24)$$

Hence we have solved d'Alembert's paradox to the extent that we now see how a force can exist on a body in a perfect fluid—by the existence of circulation around the body. But so far, for a given body in a stream, we have no way of ascertaining *how much* circulation will arise. A solution to this new paradox will be given in later paragraphs.

The lifting force Y, above, is responsible for the curved flight of ping-pong and tennis balls and, in a slightly different form, for the lift of an airplane wing. Equation (4.24) is known as the *Kutta-Joukowski theorem of lift*.

4.5. Further Discussion of $\rho V \Gamma$. Equation (4.24) is, of course, a two-dimensional relation. It may be extended to a complete three-dimensional wing with constant spanwise lift distribution by adding the wing span b. Then

$$L = \rho V \Gamma b \qquad (4.25)$$

From either Eq. (4.24) or (4.25) it is seen that the circulation Γ must increase as the velocity decreases if the lift is to remain constant.

As a matter of interest, tests have been made with rotating cylinders, and a reasonable agreement seems to exist between theory and practice. In one case (Ref. 4.3) a cylinder of 0.375 ft diameter was whirled at 3,600 rpm in a 49.2 ft per sec airstream. End conditions were such that an infinite aspect ratio was simulated, and the developed lift coefficient was 9.5.

From Eq. (4.25) we have

$$c_l = \frac{\rho V \Gamma b}{(\rho/2) c b V^2}$$

or

$$\Gamma = \frac{c V c_l}{2} \qquad (4.26)$$

For the case in question this becomes

$$\Gamma = \frac{0.375 \times 49.2 \times 9.5}{2} = 87.4 \text{ sq ft per sec}$$

as determined by the developed lift.

The theoretical circulation at the surface of the cylinder is

$$\Gamma = q_\theta \cdot 2\pi r = r\omega \cdot 2\pi r = 82.6 \text{ sq ft per sec}$$

Considering that there should probably be a slight increase in r to allow for the boundary layer, the agreement is quite close. The circulation in this example is below the critical value that would require a part of the fluid to rotate and is in the range $0.5 < q_\theta/V < 3.0$ where best agreement is found.

An important point to remember in comparing rotating circular cylinders and circular cylinders with circulation is that the tangential velocity for the former is constant, while for the latter it varies according to Eq. (4.17). Considering this condition, the agreement mentioned above is quite remarkable.

4.6. The Velocity Potential ϕ. It may be shown mathematically that the condition for irrotational motion [Eq. (4.10)] is identically equal to the condition that a function ϕ must exist such that its first partials in any direction are the velocities of the fluid that it represents in that direction.* That is,

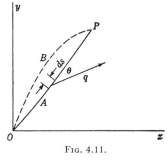

FIG. 4.11.

$$u = \frac{\partial \phi}{\partial x}, \qquad v = \frac{\partial \phi}{\partial y} \qquad (4.27)$$

$$q_r = \frac{\partial \phi}{\partial r}, \qquad q_\theta = \frac{1}{r}\frac{\partial \phi}{\partial \theta} \qquad (4.28)$$

To understand how these relations are derived, consider a point P connected to the origin by a simple curve OAP, and let ϕ be the value of the integral of the tangential component of the velocity q along OAP; that is (Fig. 4.11),

$$\phi = \int_{OAP} q \cos \theta \, ds \qquad (4.29)$$

Unlike ψ, which as long as no additional sources or sinks are involved is independent of the path selected (Fig. 3.3), the value of ϕ generally is dependent on the path. If OBP is some path other than OAP, from Sect. 4.2 the circulation

$$\Gamma = \int_{OAPBO} q \cos \theta \, ds$$
$$= \phi_{OAP} + \phi_{PBO}$$
$$= \phi_{OAP} - \phi_{OBP}$$

Now the circulation about a zone is equal to the total vortex strengths enclosed within it, and since there are none in $OAPBO$, the vorticity is zero and so is the circulation. Hence

$$\phi_{OAP} = \phi_{OBP} \qquad (4.30)$$

and it may be said that ϕ is unique. When the motion is irrotational (as above), ϕ will have a constant value at a point P regardless of the

* The function ϕ was first employed by Euler.

path followed from the origin to the point and is then called the *velocity potential.* It has the units of square feet per second.

The previous discussion concerning the effect of a body in a fluid may be extended to the case for ϕ. For instance, similarity of the equation for Γ,

$$\Gamma = \oint q \cos \theta \, ds$$

and that for ϕ [Eq. (4.29)] must have been previously noted.

If the path OAP be selected to extend around a body which has a circulation Γ, it is apparent that ϕ will be increased by Γ by the one circumlocution to which we limit ourselves.

Rewriting Eq. (4.29) using the velocity components u and v and dx and dy for ds, we have for an irrotational field:

$$\phi = \int_0^P u \, dx + v \, dy$$

or

$$d\phi = u \, dx + v \, dy \tag{4.31}$$

But from the definition of a derivative

$$d\phi = \frac{\partial \phi}{\partial x} dx + \frac{\partial \phi}{\partial y} dy$$

and equating components,

$$u = \frac{\partial \phi}{\partial x}\left(= \frac{\partial \psi}{\partial y}\right) \tag{4.32}$$

$$v = \frac{\partial \phi}{\partial y}\left(= -\frac{\partial \psi}{\partial x}\right) \tag{4.33}$$

To derive the expressions for the velocities in polar coordinates q_θ and q_r, express the components of ds as dr and $r \, d\theta$. Then, similarly,

$$d\phi = q_r \, dr + q_\theta r \, d\theta \tag{4.34}$$

$$= \frac{\partial \phi}{\partial r} dr + \frac{\partial \phi}{\partial \theta} d\theta$$

so that

$$q_r = \frac{\partial \phi}{\partial r}\left(= \frac{1}{r}\frac{\partial \psi}{\partial \theta}\right) \tag{4.35}$$

$$q_\theta = \frac{1}{r}\frac{\partial \phi}{\partial \theta}\left(= -\frac{\partial \psi}{\partial r}\right) \tag{4.36}$$

Since a flow field may be determined from ϕ, it is evident that ϕ may be used to represent a field. Hence, at least one function other than ψ exists that may represent flow patterns.

It is interesting to consider the physical meaning of ϕ to parallel the concept of flow quantity arising from ψ. In Sect. 4.9 it will be shown that equipotential lines plot perpendicular to streamlines, and the possibility that ϕ concerns pressures at once suggests itself. Now consider the units of $\rho\phi$.

$$\rho\phi = \frac{\text{lb} - \sec^2}{\text{ft}^4} \cdot \frac{\text{ft}}{\sec} = \frac{\text{lb} - \sec}{\text{sq ft}}$$

An *impulse* is a force for given time, pound-seconds, so that $\rho\phi$ must be an impulsive pressure. Thus ϕ is a function of the impulsive pressure needed to create a given flow.

4.7. Tests for Rotation and Compressibility. From Sect. 3.2 we have that incompressibility exists if

$$\frac{\partial u}{\partial x} + \frac{\partial v}{\partial y} = 0 \tag{4.37}$$

which from Eqs. (4.32) and (4.33) may be expressed

$$\frac{\partial^2\phi}{\partial x^2} + \frac{\partial^2\phi}{\partial y^2} = 0 \tag{4.38}$$

and from Eqs. (4.12) and (4.14) the motion is irrotational if

$$\frac{\partial u}{\partial y} - \frac{\partial v}{\partial x} = 0 \tag{4.39}$$

or

$$\frac{\partial^2\psi}{\partial x^2} + \frac{\partial^2\psi}{\partial y^2} = 0 \tag{4.40}$$

That is, if $\nabla^2\phi = 0$, the motion is incompressible and if $\nabla^2\psi = 0$, the motion is irrotational. Equations (4.37) and (4.39) are called the *Cauchy-Riemann equations*, and Eqs. (4.38) and (4.40) are called *Laplacians*. They lead to the reason we have been able to add stream functions or velocity potentials: in each case they have satisfied the Laplacian, and if two functions satisfy the Laplacian, their sum will too.

4.8. Determining ϕ if ψ Is Known. Stream functions, as we have defined them, exist only for incompressible flows, and it has been convenient so far to deal only with stream functions. Soon, however, we shall need the velocity potentials for each stream function so far developed. As either ϕ or ψ can be used to define a flow, it follows that, for each ψ so far developed, some ϕ corresponds. It is developed from ψ as follows.

1. Check the stream function for rotation, as unless it is irrotational no ϕ exists.

2. If irrotational, take the partials of ψ, $\dfrac{\partial \psi}{\partial x}$ and $\dfrac{\partial \psi}{\partial y}$.

3. Substitute them into the expressions for the total derivative of ϕ, using Eqs. (4.35) and (4.36).

4. Integrate the resulting differential equation according to the differential-equation laws. (One way is to integrate the terms containing x in the dx expressions holding y constant, and only the pure functions of y in the dy expression.)

Two examples follow:

Example 4.1. Find the velocity potential for $\psi = 2xy$:

1. $\nabla^2 \psi = 0$; that is, the flow is irrotational

2. $\dfrac{\partial \psi}{\partial x} = 2y ; \dfrac{\partial \psi}{\partial y} = 2x$

3. $d\phi = \dfrac{\partial \phi}{\partial x}\, dx + \dfrac{\partial \phi}{\partial y}\, dy$

$\quad = \dfrac{\partial \psi}{\partial y}\, dx - \dfrac{\partial \psi}{\partial x}\, dy$

$\quad = 2x\, dx - 2y\, dy$

4. $\phi = x^2 - y^2$

Example 4.2. Find the velocity potential for a source $\psi = Q\theta/2\pi$:

1. $\nabla^2 \psi = \dfrac{\partial^2 \psi}{\partial r^2} + \dfrac{1}{r^2}\dfrac{\partial^2 \psi}{\partial \theta^2} + \dfrac{1}{r}\dfrac{\partial \psi}{\partial r} = 0$

2. $q_r = \dfrac{1}{r}\dfrac{\partial \psi}{\partial \theta} = \dfrac{Q}{2\pi r} = \dfrac{\partial \phi}{\partial r}$

$\quad q_\theta = -\dfrac{\partial \psi}{\partial r} = 0 = \dfrac{1}{r}\dfrac{\partial \phi}{\partial \theta}$

3. $d\phi = \dfrac{\partial \phi}{\partial r}\, dr + \dfrac{\partial \phi}{\partial \theta}\, d\theta$

$\quad = \dfrac{Q}{2\pi r}\, dr + 0$

4. $\phi = \dfrac{Q}{2\pi}\ln r$

The method outlined above may be used to find the velocity potential for the basic flows of Chaps. 3 and 4, yielding

Uniform flow parallel to x axis:

$$\psi = Vy \qquad \phi = Vx \tag{4.41}$$

Source at origin:

$$\psi = \frac{Q\theta}{2\pi} \qquad \phi = \frac{Q}{2\pi}\ln r \tag{4.42}$$

Doublet at origin:

$$\psi = \frac{\mu}{2\pi}\frac{y}{r^2} \qquad \phi = -\frac{\mu}{2\pi}\frac{x}{r^2} \tag{4.43}$$

Circulation about circular cylinder of radius a at origin:

$$\psi = -\frac{\Gamma}{2\pi} \ln \frac{r}{a} \qquad \phi = \frac{\Gamma}{2\pi} \theta \qquad (4.44)$$

4.9. Plotting Equipotential Lines. Assuming equal increments of ϕ and plotting in a manner similar to the plots of streamlines, families of

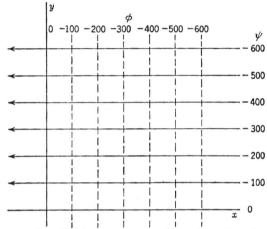

Fig. 4.12. Streamlines (solid) and equipotential lines for uniform flow to the left.

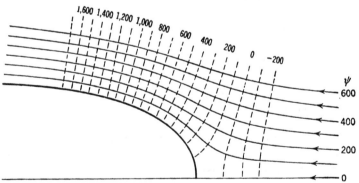

Fig. 4.13. Streamlines (solid) and equipotential lines for a source in a uniform stream. $Q = 2,000$, $V = 100$ ft per sec.

equipotential lines may be drawn. Examples of a uniform flow, a source in a uniform flow, and a doublet are shown in Figs. 4.12, 4.13, and 4.14. It is noted that the streamlines cross the equipotential lines at right angles, and hence for small increments ϕ and ψ divide the field into small squares. When larger increments are taken, the squares are distorted but the right angles remain. This is an important concept in the development of conformal transformation in the next chapter. Streamlines and equipotential lines do not form right angles at stagnation points.

Now then, we have two functions, ϕ and ψ, either of which may be used to describe a certain irrotational incompressible steady flow. Further, without manipulation the two functions plot at right angles to each other, an orthogonal system. The idea instantly presents itself that a complex variable could be formed of these functions that would also describe the flow and in addition have the mathematical characteristics of complex variables that in many cases simplify the analysis. We therefore proceed to join the velocity potential and stream function into a complex *potential function* defined as

$$w = \phi + i\psi \qquad \text{sq ft per sec} \quad (4.45)$$

for incompressible irrotational flow.

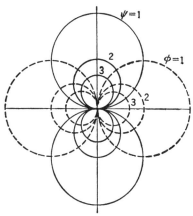

Fig. 4.14. Streamlines (solid) and equipotential lines for a doublet.

4.10. The Potential Function. Using Eq. (4.45), the flows expressed in Chap. 3 as stream functions and in Sect. 4.8 as velocity potentials may now be combined into the potential-function form. For instance, for a uniform flow parallel to the x axis, $\psi = Vy$ and $\phi = Vx$. Then

$$w = \phi + i\psi = Vx + iVy = V(x + iy) \qquad (4.46)$$

From the notation of Sect. 2.1 this becomes

$$w = Vz \qquad (4.47)$$

for a uniform flow parallel to the x axis.

In a similar manner the other basic functions become:
Uniform flow parallel to y axis:

$$w = -iVz \qquad (4.48)$$

Source at origin:

$$w = \frac{Q}{2\pi} \ln z \qquad (4.49)$$

Doublet at origin:

$$w = -\frac{\mu}{2\pi z} \tag{4.50}$$

Circular cylinder of radius a with circulation Γ:

$$w = -V\left(z + \frac{a^2}{z}\right) - \frac{i\Gamma}{2\pi}\ln\frac{z}{a} \tag{4.51}$$

We may now develop a relation for the absolute value of the velocity at some point in a flow that is described by the potential function w. The velocity will be, as written in complex notation,

$$q = u + iv \tag{4.52}$$

where u and v are the values of the horizontal and vertical velocity components. Consider the conjugate of q, $u - iv$. It has the same absolute value as $u + iv$, and hence we may write

$$|q| = |u + iv| = |u - iv| = |\sqrt{u^2 + v^2}| \tag{4.53}$$

But according to Eqs. (4.32) and (4.33) we may write $u = \dfrac{\partial \phi}{\partial x}$ and $v = -\dfrac{\partial \psi}{\partial x}$ so that

$$|q| = \left|\frac{\partial \phi}{\partial x} + i\frac{\partial \psi}{\partial x}\right| = \left|\frac{\partial(\phi + i\psi)}{\partial x}\right|$$

By Eqs. (4.45) and (2.2) this becomes

$$|q| = \left|\frac{dw}{dz}\right| \tag{4.54}$$

Equation (4.54) is used in almost every important derivation in two-dimensional airfoil development.

Before closing this introductory section on the potential function, it should be noted that neither ϕ, ψ, or w possesses any "superior" qualities as far as all flow problems are concerned. Sometimes it is advantageous to use any one of the three. The stream function ψ may be used to represent graphically a given flow; partials of ϕ in a given direction yield the velocity in that direction; and w has the advantages of complex notation.

4.11. The Blasius Equations. When the potential function w for the flow about a body is known in terms of the complex variable z, it is possible to obtain simple expressions (known as the *Blasius equations*) for the force and moment on the body. First, however, it is in order to consider how changes in pressure and momentum occur in a fluid.

Consider the application of $F = ma$ to the fluid in a control zone. The mass being acted upon is composed of different fluid particles each instant, and F should be considered as the force necessary to bring a certain mass to rest each second, or to change its velocity. Thus we usually write

$$F = \frac{\text{mass}}{\text{sec}} \cdot \Delta v \tag{4.55}$$

A simple case of analyzing the force on a body buoyed up in a vertical stream of a viscous fluid can illustrate this use of Newton's second law, a viscous fluid being selected to allow a drag force. Consider the fluid crossing the boundaries AA' and BB' (see Fig. 4.15). The change in total head between BB' and AA' may show up in one of three ways:

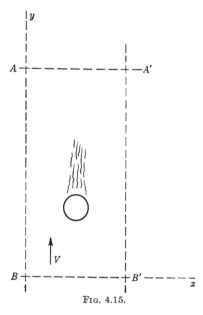

1. If the boundaries AB and $A'B'$ are parallel and solid, the law of continuity precludes any change of velocity and the force of the body on the fluid (its weight) must appear as a loss of static pressure over an area. Thus

$$-W + \int_C p \, dx = 0 \quad (4.56)$$

2. If the region is one of constant pressure (as the free atmosphere), a change of velocity and momentum

FIG. 4.15.

will occur in the wake of the body, but no pressure force. Thus

$$-W = -\int_C \rho v \, dx \cdot v \tag{4.57}$$

the force in the minus direction changing the momentum in that direction.

3. The third case is a combination of the other two where both pressure and momentum changes occur and we have

$$-W + \int_C p \, dx = -\int_C \rho v \, dx \cdot v \tag{4.58}$$

4.12. General Formula for Lift. Now let us extend Sect. 4.11 to include two-dimensional motion where both u and v velocities occur. Suppose (Fig. 4.16) that the forces on the body from the fluid are $+X$ and $+Y$, so that the forces *on the fluid* from the body will be $-X$ and $-Y$. Also, let positive momentum be along the plus axes.

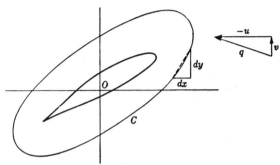

Fig. 4.16.

The pressure forces according to our setup are in the $+y$ and $-x$ directions. The momentums entering the zone have also $+y$ and $-x$ direction. Hence

$$-X - \int_C p \, dy + \rho \int_C (v \, dx - u \, dy)u = 0 \tag{4.59}$$

and

$$-Y + \int_C p \, dx + \rho \int_C (v \, dx - u \, dy)v = 0 \tag{4.60}$$

Forming $-iY$ from Eq. (4.60), we then get

$$X - iY = -\int_C p(dy + i \, dx) - \rho \int_C (u - iv)(u \, dy - v \, dx) \tag{4.61}$$

From Bernoulli's equation,

$$p = H - \tfrac{1}{2}\rho(u^2 + v^2) \tag{4.62}$$

and hence

$$X - iY = -\int_C [H - \tfrac{1}{2}\rho(u^2 + v^2)](dy + i \, dx)$$
$$- \rho \int_C (u - iv)(u \, dy - v \, dx) \tag{4.63}$$

H, being a constant, integrates to zero about a closed curve so that

$$X - iY = \tfrac{1}{2}\rho \int_C [(u^2 + v^2)(dy + i \, dx) - 2(u - iv)(u \, dy - v \, dx)] \tag{4.64}$$

Expanding and simplifying,

$$X - iY = \tfrac{1}{2}\rho i \int_C (u - iv)^2 (dx + i\, dy) \qquad (4.65)$$

but

$$\frac{dw}{dz} = u - iv; \quad z = x + iy; \quad \text{and} \quad dz = dx + i\, dy \quad (4.66)$$

so that

$$X - iY = \frac{1}{2}\rho i \int_C \left(\frac{dw}{dz}\right)^2 dz \qquad (4.67)$$

We see that the forces on a body are a function of the potential function that describes the flow about the body or, more closely, of the first derivative of the potential function. Equation (4.67) is known as the *general force equation* and will be used later to evaluate lift and drag.

4.13. General Formula for Moment. The relation between the potential function that represents the flow about a body and the moment of the body about the origin follows from the equation of angular motion. For instance (Fig. 4.16), the mass that enters the zone is $(v\, dx - u\, dy)$, and its moment of momentum is counterclockwise in the amount $\rho(v\, dx - u\, dy)(vx - uy)$. If the moment on the body is M_0, that from the body on the fluid is $-M_0$. Finally, the sum of the body moment, pressure moment, and fluid momentum is

$$-M_0 + \int_C (px\, dx + py\, dy) + \rho \int_C (v\, dx - u\, dy)(vx - uy) = 0 \quad (4.68)$$

Substituting for p from Eq. (4.62) and recalling that $\int_C H\, dz = 0$, we have

$$M_0 = -\tfrac{1}{2}\rho \int_C (u^2 + v^2)(x\, dx + y\, dy) + 2(u\, dy - v\, dx)(vx - uy) \quad (4.69)$$

$$= -\tfrac{1}{2}\rho \int_C (u^2 - v^2)(x\, dx - y\, dy) + 2uv(x\, dy + y\, dx) \quad (4.70)$$

Now consider

$$\int_C \left(\frac{dw}{dz}\right)^2 z\, dz = \int_C (u - iv)^2 (x + iy)(dx + i\, dy) \qquad (4.71)$$

Expanding and designating the real part by the abbreviation Re, we have

$$\text{Re} \int_C \left(\frac{dw}{dz}\right)^2 z\, dz = \int_C (u^2 - v^2)(x\, dx - y\, dy) + 2uv(x\, dy + y\, dx)$$

$$(4.72)$$

From Eq. (4.70) we now see that

$$M_0 = -\frac{1}{2}\rho\,\mathrm{Re}\int_C \left(\frac{dw}{dz}\right)^2 z\,dz \qquad \text{ft lb per ft of span} \qquad (4.73)$$

and the moment about the origin on a body in a moving stream also is a function of the potential function that describes the flow about the body. Equations (4.67) and (4.73) are known as the *Blasius equations* and were presented by Blasius in 1908.

4.14. Summary of Important Relations. It seems in order to organize and state in one place the important relations so far derived.

Rectilinear Coordinates   *Polar Coordinates*

1. Velocity relations:

$$u = \frac{\partial\psi}{\partial y} = \frac{\partial\phi}{\partial x} \qquad\qquad q_\theta = -\frac{\partial\psi}{\partial r} = \frac{1}{r}\frac{\partial\phi}{\partial\theta}$$

$$v = -\frac{\partial\psi}{\partial x} = \frac{\partial\phi}{\partial y} \qquad\qquad q_r = \frac{1}{r}\frac{\partial\psi}{\partial\theta} = \frac{\partial\phi}{\partial r}$$

$$|q| = \sqrt{u^2 + v^2} = \left|\frac{dw}{dz}\right| \qquad |q| = \sqrt{q_r{}^2 + q_\theta{}^2} = \left|\frac{dw}{dz}\right|$$

2. Conditions for incompressible flow:

$$\frac{\partial u}{\partial x} + \frac{\partial v}{\partial y} = 0 \qquad\qquad \frac{\partial q_r}{\partial r} + \frac{q_r}{r} + \frac{1}{r}\frac{\partial q_\theta}{\partial\theta} = 0$$

$$\frac{\partial^2\phi}{\partial x^2} + \frac{\partial^2\phi}{\partial y^2} = 0 \qquad \frac{\partial^2\phi}{\partial r^2} + \frac{1}{r}\frac{\partial^2\phi}{\partial\theta^2} + \frac{1}{r}\frac{\partial\phi}{\partial r} = 0$$

$$\nabla\cdot\mathbf{q} = 0; \qquad \psi \text{ exists, } w \text{ exists}$$

3. Conditions for irrotational flow:

$$\frac{\partial u}{\partial y} - \frac{\partial v}{\partial x} = 0 \qquad\qquad \frac{1}{r}q_\theta + \frac{\partial q_\theta}{\partial r} - \frac{1}{r}\frac{\partial q_r}{\partial\theta} = 0$$

$$\frac{\partial^2\psi}{\partial x^2} + \frac{\partial^2\psi}{\partial y^2} = 0 \qquad \frac{1}{r}\frac{\partial\psi}{\partial r} + \frac{\partial^2\psi}{\partial r^2} + \frac{1}{r^2}\frac{\partial^2\psi}{\partial\theta^2} = 0$$

$$\nabla\times\mathbf{q} = 0; \qquad \phi \text{ exists, } w \text{ exists}$$

4. Expressions concerning uniform flow V, to the left:

$$u = -V$$
$$\psi = -Vy \qquad \psi = -Vr\sin\theta$$
$$\phi = -Vx \qquad \phi = -Vr\cos\theta$$
$$w = -Vz \qquad w = -Vre^{i\theta}$$

5. Expressions concerning source of quantity Q at origin:

$$u = \frac{Q}{2\pi} \frac{x}{x^2 + y^2} \qquad q_r = \frac{Q}{2\pi r}$$

$$v = -\frac{Q}{2\pi} \frac{y}{x^2 + y^2} \qquad q_\theta = 0$$

$$\psi = \frac{Q}{2\pi} \tan^{-1} \frac{y}{x} \qquad \psi = \frac{Q\theta}{2\pi}$$

$$\phi = \frac{Q}{2\pi} \ln (x^2 + y^2)^{\frac{1}{2}} \qquad \phi = \frac{Q}{2\pi} \ln r$$

$$w = \frac{Q}{2\pi} \ln z \qquad w = \frac{Q}{2\pi} (\ln r + i\theta)$$

6. Expressions concerning doublet of strength μ at origin:

$$u = \frac{\mu(x^2 - y^2)}{2\pi(x^2 + y^2)^2} \qquad q_r = -\frac{\mu}{2\pi} \frac{\cos\theta}{r^2}$$

$$v = \frac{\mu xy}{\pi(x^2 + y^2)^2} \qquad q_\theta = \frac{\mu}{2\pi} \frac{\sin\theta}{r^2}$$

$$\psi = \frac{\mu y}{2\pi(x^2 + y^2)} \qquad \psi = \frac{\mu}{2\pi} \frac{\sin\theta}{r}$$

$$\phi = -\frac{\mu x}{2\pi(x^2 + y^2)} \qquad \phi = -\frac{\mu}{2\pi} \frac{\cos\theta}{r}$$

$$w = -\frac{\mu}{2\pi z} \qquad w = -\frac{\mu}{2\pi r e^{i\theta}}$$

7. Expressions concerning vortex of strength Γ at origin:

$$u = -\frac{\Gamma}{2\pi} \frac{y}{x^2 + y^2} \qquad q_r = 0$$

$$v = \frac{\Gamma}{2\pi} \frac{x}{x^2 + y^2} \qquad q_\theta = \frac{\Gamma}{2\pi r}$$

$$\psi = -\frac{\Gamma}{4\pi} \ln \frac{(x^2 + y^2)^{\frac{1}{2}}}{a} \qquad \psi = -\frac{\Gamma}{2\pi} \ln \frac{r}{a}$$

$$\phi = \frac{\Gamma}{2\pi} \tan^{-1} \frac{y}{x} \qquad \phi = \frac{\Gamma}{2\pi} \theta$$

$$w = -\frac{i\Gamma}{2\pi} \ln \frac{z}{a} \qquad w = -\frac{i\Gamma}{2\pi} (\ln r + i\theta - \ln a)$$

PROBLEMS

4.1. Check the determination of ϕ in Eqs. (4.41) to (4.44).

4.2. Check the determination of w in Eqs. (4.48) to (4.51).

4.3. Sketch the velocity distribution in a real vortex. What is the relation between velocity and radius in the irrotational part of the field? The rotational part?

4.4. If $\phi = x^2 - y^2$, find ψ.

4.5. If $\psi = x^2 - y^2$, find ϕ.

4.6. If $\phi = x^2 - y^2$, find the total velocity at $(2, 4)$.

4.7. Does the potential function exist for irrotational flow only?

4.8. If $\psi = ky/2\pi r^2$, find ϕ and w.

4.9. A circular cylinder, 2 ft diameter, is in a uniform stream of 100 ft per sec. $\Gamma = 300\pi$. Find the location of the stagnation points and the largest negative pressure.

4.10. For Prob. 4.9, find the flow direction and magnitude at $r = 2$, $\theta = 30°$.

4.11. Discuss the rotational part of the perfect-fluid vortex

$$\psi = -\frac{\Gamma}{2\pi} \ln \frac{r}{a}$$

REFERENCES

4.1. W. F. Durand, "Aerodynamic Theory," Vol. I, pp. 130–139, Verlag Julius Springer, Berlin, 1934.

4.2. H. Glauert, "Aerofoil and Airscrew Theory," pp. 33–49, Cambridge, University Press, London, 1926.

4.3. E. G. Reid, Tests of Rotating Cylinders, *TN* 209, 1924.

4.4. N. A. V. Piercy, "Aerodynamics," pp. 137–166, English University Press, London, 1947.

4.5. S. Goldstein, "Modern Developments in Fluid Dynamics," Oxford University Press, New York, 1938.

CHAPTER 5

THE JOUKOWSKI TRANSFORMATION

5.1. Conformal Transformation. It is possible to have figures of very different shape related to each other in some manner. An example might be shadows of one object as cast on a wall or on a floor. The shadows are obviously related, although entirely different in shape. So it is with many pairs of figures. While a simple factor might make one square four times another, a more complex relation might warp the square into a rhombus or to a rectangle.

For instance, consider a figure (see Fig. 5.1) described in the z plane by $z = x + iy$ and another figure in the ζ plane described by $\zeta = \xi + i\eta$. Further, let the two figures be related by $\zeta = f(z)$. An infinitesimal line element in the z plane dz becomes an infinitesimal element in the ζ plane $d\zeta$ according to

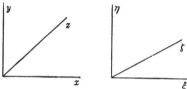

FIG. 5.1. Argand diagrams for the z and ζ planes.

$$d\zeta = \frac{\partial \zeta}{\partial z}\, dz$$

Now both $d\zeta$ and dz are complex and may be changed in length or rotated (see Sect. 2.1), but as long as $\partial\zeta/\partial z$ has one value and only one value at each point, the infinitesimal zone about point P_z will remain similar to that about P_ζ and the transformation is said to be *conformal*. The shape of large figures in the z and ζ planes may be different because $\partial\zeta/\partial z$ may have different values at different points. The streamlines and equipotential lines which cross at right angles in the w plane ($w = \phi + i\psi$) will continue to cross at right angles if the w plane is conformally transformed—except at singularities where $\dfrac{d\zeta}{dz}$ is either zero or infinite. Such points require special treatment.

To consider a simple case, let a point in the z plane be represented by $x + iy$ and one in the ζ plane by $\xi + i\eta$. Further let

$$\zeta = z^2 \tag{5.1}$$

Then

$$\xi + i\eta = (x + iy)^2 = x^2 - y^2 + 2ixy$$

or

$$\xi = x^2 - y^2$$
$$\eta = 2xy$$

Consider, for instance, lines of x = const in the z plane which plot as vertical lines shown in Fig. 5.2a.

In the ζ plane, when $x = 1$, $\eta = 2y$ and $\xi = 1 - y^2 = 1 - (\eta^2/4)$, so that

$$\eta = \sqrt{4(1 - \xi)} \tag{5.2}$$

Thus the straight lines in the z plane have been transformed into parabolas in the ζ plane by the transformation $\zeta = z^2$ (see Fig. 5.2b).

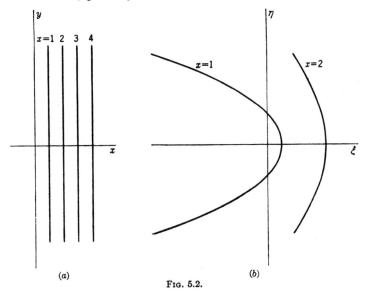

(a)   (b)

Fɪɢ. 5.2.

5.2. The Joukowski Transformation.   The equation of the flow about a circular cylinder with circulation has already been developed, and it has been shown that the solution yields a lift force.   If a transformation could be found that would transform that flow to the flow about an airfoil (the circle becoming the airfoil), then we could, in a manner similar to the treatment of the circle, find the pressure distribution around the airfoil and its lift.

The first thought that comes to mind regarding the above is to seek a function that reduces y values only in some orderly manner that changes a circle to an airfoil shape of chord equal to the diameter of the circle. However, while such a function could doubtless be found, the solution advanced [apparently simultaneously and independently by Kutta (born

1867) in Germany and Joukowski (1847–1921) in Russia] operates on both x and y values, moving points down and away from the y axis.

The Kutta-Joukowski transformation is

$$\zeta = z + \frac{c_1^2}{z} \tag{5.3}$$

where c_1 is a constant. Properly employed, this transformation will transform a circle in the z plane into an airfoil in the ζ plane.

The mechanism of the transformation is shown by the following graphical interpretation:

If $z = re^{i\theta}$ and $\zeta = \xi + i\eta$, then the Joukowski transformation may be written

$$\xi + i\eta = re^{i\theta} + \frac{c_1^2}{r} e^{-i\theta}$$

Equating reals and imaginaries

$$\xi = \left(r + \frac{c_1^2}{r}\right) \cos \theta \tag{5.4}$$

$$\eta = \left(r - \frac{c_1^2}{r}\right) \sin \theta \tag{5.5}$$

Consider Fig. 5.3, in which the transformed point A' in the ζ plane is plotted on the same axes as the original point A in the z plane. Expanding Eqs. (5.4) and (5.5), we have

$$\xi = r \cos \theta + \frac{c_1^2}{r} \cos \theta \tag{5.6}$$

and

$$\eta = r \sin \theta - \frac{c_1^2}{r} \sin \theta \tag{5.7}$$

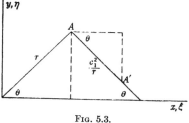

Fig. 5.3.

Examination of Fig. 5.3 discloses that the movement of point A is in accordance with the Joukowski transformation; points in the first quadrant are moved down and to the right, while points in the second quadrant move down and to the left.

Further study is required to see why the Kutta-Joukowski transformation makes an airfoil and not some other flat oval. Indeed, special care is needed to get an airfoil.

Let the circle have its center to the right of the y axis (Fig. 5.4). Then, for a given value of $\theta_1 = \theta_2$, $r_1 > r_2$ and the quantity $\left(r_1 - \frac{c_1^2}{r_1}\right)$

$> \left(r_2 - \dfrac{c_1{}^2}{r_2} \right)$. Thus from Eq. (5.7) the ordinate of A' is greater than that of B', and an airfoil shape develops. The location of the center of the circle is a prime factor in obtaining the airfoil. It will be shown

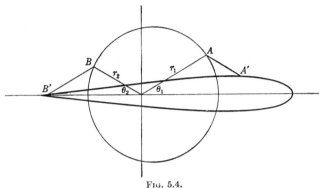

Fig. 5.4.

later that with the center at $(0, 0)$ a straight-line or flat-plate airfoil is obtained, and with it on the y axis a circular-arc airfoil results.*

As mentioned previously, special attention must be given to points at which $d\zeta/dz$ becomes zero or infinite. The derivative of the Joukowski transformation becomes negatively infinite at the origin, but since that point will be inside our body, it will not appear in the flow and will not be of interest. The zero values which appear at $\pm c_1$ in the z plane will receive special consideration.

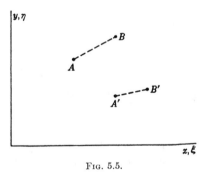

Fig. 5.5.

5.3. Transformation of the Velocity.
From Fig. 5.3, the Joukowski transformation moves a point A in the first quadrant down and to the right to point A', and in a similar manner a point B would be moved to B' (Fig. 5.5). The same amount of fluid that passes between A and B in the z plane must after transformation pass between A' and B' in the ζ plane. The velocities in the two planes are then inversely proportional to the distance between two points; or

---

* A simple mechanical device for rapidly performing the Joukowski transformation is described by Bairstow in Ref. 5.3.

$$|q_\zeta| = \frac{|q_z|}{\left|\dfrac{A'B'}{AB}\right|} \tag{5.8}$$

The important principle for us to note here is that, since the conformal transformation increases velocities as it decreases distances, the *circulation around a body will remain constant through a transformation.* We shall need this principle in the next chapter.

When AB is taken as very small and equal to dz, $A'B'$ will equal $d\zeta$. Hence

$$|q_\zeta| = \frac{|q_z|}{\left|\dfrac{d\zeta}{dz}\right|} = \frac{|q_z|}{\left|1 - \dfrac{c_1^2}{z^2}\right|} \tag{5.9}$$

To obtain the actual value of $\dfrac{d\zeta}{dz}$, recall that the absolute value of a complex variable equals the square root of the sum of the squares of the real and imaginary parts. Hence, separating z in Eq. (5.9) into real and imaginary parts, we have

$$\left|\frac{d\zeta}{dz}\right| = \left|\sqrt{\left(1 - \frac{c_1^2}{r^2}\cos 2\theta\right)^2 + \left(\frac{c_1^2}{r^2}\sin 2\theta\right)^2}\right| \tag{5.10}$$

We shall leave the discussion of Eq. (5.10) to a later time.

5.4. Lift of a Joukowski Airfoil. In Chap. 4 relations were developed that enable the lift and moment on a body in a fluid to be found provided that the potential function for the flow about the body is known. We now proceed to find the potential function (or, more accurately, its first derivative) of a Joukowski airfoil. While it is true that we know the potential function for a circular cylinder with circulation [Eq. (4.51)] and a transformation that will transform the circular cylinder into an airfoil, several interim steps remain, all of considerable importance.

To recapitulate, we have seen how lift develops when a circulation arises, and it is common knowledge that lift is also developed by an airfoil. But we cannot transform a circular cylinder with circulation and lift to an airfoil at some angle α until we tie together the angle of attack of the airfoil and its lift, and the lift (and circulation) of the cylinder.

In other words, we need the condition that will define the airfoil angle of attack for the assumed circulation. The answer, given first by Kutta, is that "enough circulation will arise to permit the flow to leave the trailing edge smoothly."* Thus the greatest importance in all theoretical works is attached to the angle at the trailing edge.

---

* A "smooth" flow in perfect-fluid theory means one without infinite velocities.

In practice, an abrupt change of camber near the trailing edge would in all probability produce separation rather than a large change of lift, but in another sense there is always some mean effective angle at the trailing edge which does determine the lift.

To return to the transformation problem, several preceding steps hint the necessary approach. For instance, Eq. (5.9) tells us that, if $d\zeta/dz = 0$ at a point which later transforms to a point on the airfoil, the velocity in the transformed plane will be infinite unless the velocity at the corresponding point on the circular cylinder is zero. It behooves us then to make the point that later becomes the airfoil trailing edge, a stagnation point. It will then transform to a stagnation point if $d\zeta/dz \neq 0$ at that point for the particular transformation employed. (There are many others besides the Joukowski.) If $d\zeta/dz = 0$ at the stagnation point on

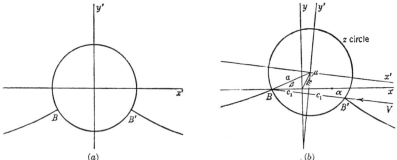

(a)   .(b)
Fig. 5.6. The circular cylinder as originally derived (a), and after being moved to the location necessary for the satisfactory use of the transformation (b).

the circle, the value of q_ζ cannot be found by Eq. (5.9) but at least it can be shown to be finite or zero.

Now then, from Eq. (5.6), points in the z plane that are on the x axis ($\theta = 0$, π) will be on the ξ axis in the ζ plane. Since (as will later be shown) the trailing edge of the airfoil is on the ξ axis, it becomes apparent that to make the airfoil trailing edge correspond to a stagnation point on the circular cylinder the cylinder must be moved up before transforming so that the stagnation points (lowered by the effect of circulation as shown in Fig. 4.9) will be on the x axis. Unless this procedure is followed, the rear stagnation point on the circular cylinder would not transform to the trailing edge of the airfoil.

Another reason for selecting the rear stagnation point of the circle at $(-c_1)$ is that the nature of the transformation makes such points into a cusp, which, although overdoing the sharp-trailing-edge requirement a trifle, at least makes a reasonable airfoil. Selecting our circle with B

inside by a small amount gives a finite trailing-edge angle, but the method of then determining the circulation through Kutta's condition becomes lost.

A further condition for the circle that is to be transformed is that its center must be to the right of the origin in order that the transformation will produce a reasonable airfoil. Thus we see that we shall want to transform a special z circle with a center at μ and radius a that meets the above conditions. An additional advantage accrues from letting the z circle enclose the singular point at $+c_1$ so that it need not be considered on the airfoil.

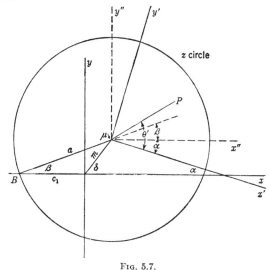

FIG. 5.7.

To introduce the concept of an angle of attack, we shall finally rotate the axes through an angle α, yielding the z circle with radius a on the $x'y'$ axes in Fig. 5.7.

Now we already have the equation of a circular cylinder with circulation [Eq. (4.51)], but it was derived relative to x, y axes when the free stream was along x. With our new setup, the free stream is along x', and hence Eq. (4.51) as we now have it lies in the $z' = x' + iy'$ plane. It becomes

$$w = -V\left(z' + \frac{a^2}{z'}\right) - i\frac{\Gamma}{2\pi}\ln\frac{z'}{a} \tag{5.11}$$

and

$$\frac{dw}{dz'} = -V\left(1 - \frac{a^2}{(z')^2}\right) - \frac{i\Gamma}{2\pi z'} \tag{5.12}$$

As we have previously explained, the point on the circular cylinder selected to become the airfoil trailing edge must be a stagnation point. Thus the value of Γ (which is the same for both cylinder and airfoil) must be such that the velocity

$$|q| = \left|\frac{dw}{dz'}\right| = 0 \tag{5.13}$$

at B, the stagnation point, where

$$z_B' = -ae^{i(\alpha+\beta)} \tag{5.14}$$

Then

$$\frac{dw}{dz'} = -V\left(1 - \frac{a^2}{a^2e^{2i(\alpha+\beta)}}\right) - \frac{i\Gamma}{2\pi(-ae^{i(\alpha+\beta)})} = 0 \tag{5.15}$$

and

$$i\Gamma = 2\pi aV(e^{i(\alpha+\beta)} - e^{-i(\alpha+\beta)}) \tag{5.16}$$

By Eq. (2.34)

$$\Gamma = 4\pi aV \sin(\alpha + \beta) \tag{5.17}$$

This very important relation ties in the circulation with the airfoil angle of attack and its camber, which, as we shall see, is controlled by β.

The airfoil is derived from the circle by the Joukowski transformation.

$$\zeta = z + \frac{c_1^2}{z}$$

And to determine the forces and moments in the ζ plane, we must find $dw/d\zeta$ to use instead of dw/dz in Eqs. (4.68) and (4.74).

The general point P is located in the $x'y'$ plane by

$$z_P' = r'e^{i\theta'} \tag{5.18}$$

or, by Eq. (2.5), at

$$z_P'' = z_P'e^{-i\alpha}$$

But

$$z_P = z_P'' + me^{i\delta} = z_P'e^{-i\alpha} + me^{i\delta}$$

so that (omitting the subscript)

$$z' = (z - me^{i\delta})e^{i\alpha} \tag{5.19}$$

and

$$\frac{dz'}{dz} = e^{i\alpha} \tag{5.20}$$

We may now evaluate $dw/d\zeta$ by the relation

$$\frac{dw}{d\zeta} = \frac{dw}{dz'}\frac{dz'}{dz}\frac{dz}{d\zeta} \tag{5.21}$$

$$\frac{dw}{d\zeta} = \left(-V + \frac{Va^2}{(z')^2} - \frac{i\Gamma}{2\pi z'}\right)(e^{i\alpha})\left(1 - \frac{c_1^2}{z^2}\right)^{-1}$$

Substituting for z' from Eq. (5.19), multiplying by the term $e^{i\alpha}$, and expanding terms in the denominator by the binomial theorem, we have

$$\frac{dw}{d\zeta} = \left[-Ve^{i\alpha} + Va^2e^{-i\alpha} \left(z^{-2} + 2z^{-3}me^{i\delta} + \cdots \right) \right.$$
$$\left. + \frac{i\Gamma}{2\pi} \left(z^{-'} + z^{-2}me^{i\delta} \right) \right] \left[1 + \frac{c_1^2}{z^2} \right]$$
$$= -Ve^{i\alpha} + \frac{Va^2e^{-i\alpha}}{z^2} - \frac{i\Gamma}{2\pi z} - \frac{i\Gamma me^{i\delta}}{2\pi z^2} - \frac{Ve^{i\alpha}c_1^2}{z^2} \tag{5.22}$$

Since Cauchy's theorem is to be used shortly, we shall need in the final step only terms as high as $1/z$. At this point, however, we have kept terms to $1/z^2$ to allow for multiplying through by z in the moment equation.

Squaring and keeping only terms below $1/z^2$, we have

$$\left(\frac{dw}{d\zeta} \right)^2 = V^2e^{2i\alpha} + \frac{iV\Gamma e^{i\alpha}}{\pi z} - \frac{2a^2V^2}{z^2} + \frac{iV\Gamma e^{i\alpha}me^{i\delta}}{\pi z^2} + \frac{2V^2c_1^2e^{2i\alpha}}{z^2} - \frac{\Gamma^2}{4\pi^2z^2} \tag{5.23}$$

The general force equation

$$X - iY = \frac{1}{2}\rho i \oint \left(\frac{dw}{d\zeta} \right)^2 \frac{d\zeta}{dz} dz \tag{5.24}$$

Substituting from Eqs. (5.9) and (5.23), and integrating by Cauchy's theorem, we get

$$X - iY = \frac{1}{2}\rho i (2\pi i) \frac{iV\Gamma e^{i\alpha}}{\pi}$$
$$= -i\rho V\Gamma e^{i\alpha} = -i\rho V\Gamma (\cos \alpha + i \sin \alpha)$$

from which

$$X = \rho V\Gamma \sin \alpha \tag{5.25}$$
$$Y = \rho V\Gamma \cos \alpha \tag{5.26}$$

These forces, however, are along the x and y axes and hence do *not* represent lift and drag. They do represent, when summed perpendicular and parallel to the wind direction, a force $\rho V\Gamma$ perpendicular to the free airstream (the lift) and zero force parallel to it. Thus in potential two-dimensional flow (negligible viscosity) there is no drag.

Substituting for the circulation from Eq. (5.7), we have

$$L = \frac{\rho}{2} V^2 \cdot 4a \cdot 2\pi \sin (\alpha + \beta) \tag{5.27}$$

which is familiar in the form

$$L = \frac{\rho}{2} V^2 S c_l$$

This equation, stating that the lift on an airfoil is proportional to the dynamic pressure, the area (approximately $4a$), and the angle of attack as measured from zero lift, is one of the most fundamental and important so far derived.

We should note two items from Eq. (5.27). The first is that the lift curve is a straight line only to the extent that $\sin (\alpha + \beta)$ is "straight"— as it happens, a pretty good approximation for the range of 0 to 10°; second, that the value at which $\sin (\alpha + \beta)$ would yield a maximum lift ($\alpha \cong 90°$) has no practical application since viscous effects produce a "stall" long before 90° α is even approached.

Equation (5.27) has in addition an interesting historical significance. Newton's development of the force on a body in a "free" fluid showed that the lift would vary as the sine squared of its angle of attack. Others, disregarding the fact that Newton was *then* discussing what amounted to rarefied air, concluded that flying machines were not possible, and it has been estimated that the "damper" thus imposed on would-be aeronauts postponed actual flight by perhaps a century. It was not until 1790 that the sine-squared law was thoroughly discredited or, better, reexamined.

A modern note is introduced by the work of Ivey, Klunker, and Bowen (Ref. 5.5), who show that, at very high supersonic speeds, the lift of a two-dimensional flat plate *is* proportional to the square of the angle of attack.

5.5. Moment of a Joukowski Airfoil. It has previously been shown that, if w describes the flow about a body, its moment about the origin

$$M_0 = -\frac{1}{2}\rho \operatorname{Re} \int_C \left(\frac{dw}{d\zeta}\right)^2 \zeta \, d\zeta \tag{5.28}$$

Substituting from Eq. (5.23) and writing

$$d\zeta = \frac{d\zeta}{dz} dz$$

we have

$$M_0 = -\frac{1}{2}\rho \operatorname{Re} \int_C \left(V^2 e^{2i\alpha} + \frac{iV\Gamma e^{i\alpha}}{\pi z} - \frac{2a^2 V^2}{z^2} + \frac{iV\Gamma e^{i\alpha} m e^{i\delta}}{\pi z^2} \right.$$
$$\left. + \frac{2V^2 e^{2i\alpha} c_1^2}{z^2} - \frac{\Gamma}{4\pi^2 z^2} \right) \left(z + \frac{c_1^2}{z} \right) \left(1 - \frac{c_1^2}{z^2} \right) dz$$

The last two terms multiply to $\left(z - \frac{c_1^4}{z^3} \right)$, and hence, completing the rest of the multiplying (disregarding terms other than $1/z$ as equal to 0 by Cauchy's second theorem),

$$M_0 = -\frac{1}{2}\rho \operatorname{Re}(2\pi i)\left[\frac{iV\Gamma m e^{i(\alpha+\delta)}}{\pi} + 2V^2 e^{2i\alpha}c_1{}^2 - 2a^2 V^2 - \frac{\Gamma^2}{4\pi^2}\right]$$

$$= \rho V\Gamma m \cos(\alpha + \delta) + 2\rho\pi V^2 c_1{}^2 \sin 2\alpha \quad (5.29)$$

This is the moment about the origin. The moment about the center of the circle becomes (see Fig. 5.8)

$$M_\mu = M_0 - Lm\cos(\alpha + \delta)$$

So, from Eq. (5.29),

$$M_\mu = 2\pi\rho V^2 c_1{}^2 \sin 2\alpha + \rho V\Gamma m$$
$$\cos(\alpha + \delta) - Lm\cos(\alpha + \delta)$$
$$= \frac{\rho}{2}V^2 4\pi c_1{}^2 \sin 2\alpha \quad (5.30)$$

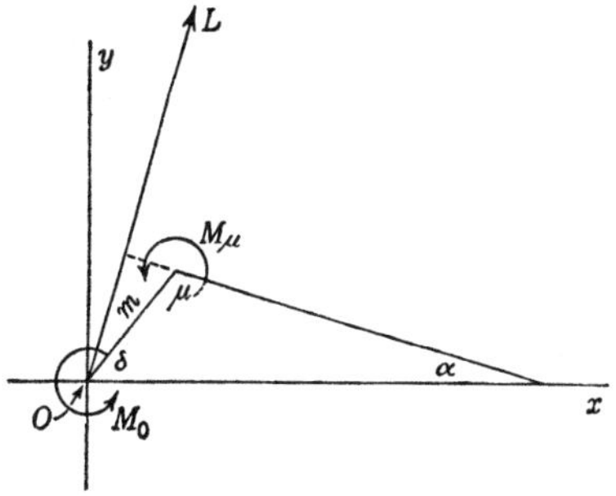

5.6. General Lift and Moment Coefficients. It is now in order to

FIG. 5.8.

reduce the lift and moment formulas to more convenient forms. We shall assume the airfoil chord $\cong 4c_1 \cong 4a$, and *all conclusions in this section will be to that approximation.*

The lift is given by Eq. (5.27) as

$$L = \frac{\rho}{2}V^2 \cdot 4a \cdot 2\pi \sin(\alpha + \beta)$$

and it is seen that the angle of zero lift is $-\beta$,

$$c_l \cong 2\pi\sin(\alpha + \beta) \quad (5.31)$$

and

$$\frac{dc_l}{d\alpha} \cong 2\pi \text{ per radian} \quad (5.32)$$

Thus the *slope of the lift curve is independent of the camber.*

To find the moment coefficient as used in practice (we rarely know either the original origin or the circle's center), we note that the point μ transforms almost exactly to the mid-point of the airfoil.

Hence the moment coefficient about the half chord is

$$c_{m\frac{1}{2}} = \frac{(\rho/2)V^2 \cdot c_1{}^2 \cdot 4\pi \sin 2\alpha}{(\rho/2)V^2 \cdot (4c_1{}^2)^2} = \frac{\pi}{2}\alpha \quad (5.33)$$

which demonstrates that the *moment coefficient about the half chord is independent of the camber.*

Transferring the above moment to the quarter chord, we have

$$c_{m\frac{1}{4}} = c_{m\frac{1}{2}} - \frac{c_l}{4}$$

$$= -\frac{\pi}{2}\beta \qquad (5.34)$$

which demonstrates that *the moment coefficient about the quarter chord is independent of the angle of attack (or c_l)* and its value is a function of the camber.

Thus the quarter chord, to a first approximation, is the *aerodynamic center* of a Joukowski airfoil where the aerodynamic center is defined as the point about which the moment coefficient is constant with changes in angle of attack. Reference to Table 7.6 indicates that this conclusion may also be extended to all airfoils with first-order accuracy.

The conclusion of Eq. (5.33) that the moment coefficient is constant about the half chord with changes in camber is strikingly validated by wind-tunnel tests of two-dimensional airfoils mounted at the half chord. Even though a flap be lowered, the moment coefficient remains practically unchanged at a constant angle of attack.

5.7. General Center-of-pressure Equation. The center of pressure (C.P.), defined as the point at which the resultant force vector intersects the chord line, may be found by writing

$$c_{mle} = c_{m\frac{1}{4}} - \frac{c_l}{4} \qquad (5.35)$$

and, assuming that all of the moment is due to the lift so that

$$c_{mle} = (\text{C.P.})\, c_l \qquad (5.36)$$

we get, substituting,

$$\text{C.P.} = \frac{c_{m\frac{1}{4}}}{c_l} - \frac{1}{4} \, \cdot \qquad (5.37)$$

The above equation leads to the interesting conclusion that if the moment about the quarter chord is large there will be a large center-of-pressure travel. If $c_{m\frac{1}{4}}$ is zero, the center of pressure is constant and at the quarter chord. (The minus sign comes from the assumption that a plus lift produces a minus moment about the leading edge.)

Some years ago the center of pressure was a much used parameter, and airfoils were divided into those with stable-center-of-pressure travel (the center of pressure moves rearward with increasing angle of attack), neutral-center-of-pressure travel, and unstable-center-of-pressure travel. But although the behavior of the center of pressure is interesting, a stability study is more simply made using the location of the aerodynamic center and the moment coefficient about it instead of the center of pressure and its position. Currently, the center-of-pressure method of analysis seems to be disappearing from the literature.

5.8. The General Airfoil Equation. The general equations for the Joukowski transformation [Eqs. (5.6) and (5.7)] may be combined to yield a useful relation, as follows:

Square both equations, and multiply Eq. (5.6) by $\sin^2 \theta$ and Eq. (5.7) by $\cos^2 \theta$. Then subtract the second from the first to get

$$\xi^2 \sin^2 \theta - \eta^2 \cos^2 \theta = 4c_1{}^2 \sin^2 \theta \cos^2 \theta \qquad (5.38)$$

This relation is called the *general airfoil equation.* Its use will be demonstrated later.

We may now apply the general lift, moment, and center-of-pressure relations to a number of Joukowski airfoils obtained by varying the center of the z circle and making minor changes to the transformation. The procedure will be the same in each case:

1. Derive relations between a, c_1, and r from the geometry of the case in question.

2. Substitute the above into the Joukowski transformation.

3. Separate out ξ and η, and analyze the shape they represent.

4. Substitute into the general lift and moment equations to get the theoretical characteristics of the particular airfoil.

5.9. The Flat-plate Airfoil. Consider the case when the z circle has its center at the origin (see Fig. 5.9). Then $r = c_1 = a$, and $\beta = 0$. The transformation becomes

$$\zeta = z + \frac{c_1{}^2}{z} = re^{i\theta} + \frac{c_1{}^2}{r} e^{i\theta}$$

$$= 2c_1 \cos \theta$$

From $\zeta = \xi + i\eta$,

$$\xi = 2c_1 \cos \theta \qquad (5.39)$$

and

$$\eta = 0 \qquad (5.40)$$

Equations (5.39) and (5.40) describe the airfoil and show that the circle has been transformed into a flat-plate airfoil with a chord from $+2c_1$ to $-2c_1$ as θ varies from 0 to π. The total chord $= 4c_1$ (exactly). Substituting the above conditions into Eq. (5.27), we have

$$L = \frac{\rho}{2} V^2 \cdot 4c_1 \cdot 2\pi\alpha$$

$$c_l = 2\pi\alpha \text{ (exactly)} \qquad (5.41)$$

$$\frac{dc_l}{d\alpha} = 2\pi \text{ per radian (exactly)} \qquad (5.42)$$

From $\alpha_{ZL} = -\beta$ we have

$$\alpha_{ZL} = 0 \tag{5.43}$$

and from Eqs. (5.34) and (5.37)

$$c_{m_{\frac{1}{4}}} = 0 \tag{5.44}$$

and

$$\text{C.P.} = -\tfrac{1}{4}$$

The above conclusions as to lift and moment of the flat-plate airfoil are not in good agreement with practice as viscous effects are not small

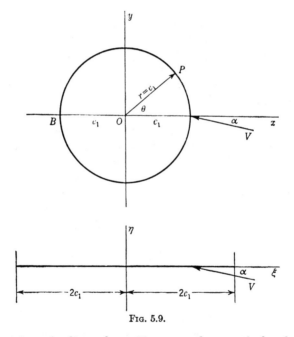

FIG. 5.9.

at the too sharp leading edge. However, the practical value of this theoretical study later becomes evident when it is shown that the pressure distribution gained with angle of attack by *any* airfoil is very nearly that developed by a flat plate at that angle.

The fact that the center of pressure was not at the half chord (as a layman might guess) was noted experimentally by Avanzine in 1804.

5.10. The Circular-arc Airfoil. When the center of the z circle is placed on the y axis, the Joukowski transformation yields a thin circular-arc airfoil.

From Fig. 5.10 we have, using rectangular coordinates to represent the circle,

$$x^2 + (y - m)^2 = a^2 \qquad (5.45)$$

or

$$r^2 \cos^2 \theta + r^2 \sin^2 \theta - 2rm \sin \theta + m^2 = a^2$$

Since $a^2 = m^2 + c_1^2$, the above equation becomes

$$r - \frac{c_1^2}{r} = 2m \sin \theta$$

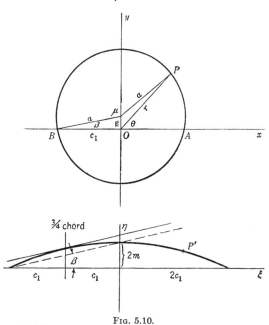

FIG. 5.10.

which by Eq. (5.5) is

$$\eta = 2m \sin^2 \theta \qquad (5.46)$$

Substituting Eq. (5.46) into the general airfoil equation [Eq. (5.38)] and dividing both sides by $\eta/2m$, we get

$$\xi^2 + \eta^2 + 2c_1 \left(\frac{c_1}{m} - \frac{m}{c_1}\right) \eta = 4c_1^2$$

which, by completing the square by adding $c_1^2 \left(\frac{c_1}{m} - \frac{m}{c_1}\right)^2$ to both sides, yields

$$\xi^2 + \left[\eta + c_1 \left(\frac{c_1}{m} - \frac{m}{c_1}\right)\right]^2 = c_1^2 \left[4 + \left(\frac{c_1}{m} - \frac{m}{c_1}\right)^2\right] \qquad (5.47)$$

This is the equation of the airfoil and also of a circle of general type $x^2 + (y + h)^2 = k^2$. Its center is at $-c_1 \left(\dfrac{c_1}{m} - \dfrac{m}{c_1} \right)$, and it passes through $\eta = 0$ at, solving,

$$\xi = +2c_1, \; -2c_1 \tag{5.48}$$

From Eq. (5.47) the maximum ordinate occurs when $\xi = 0$, which corresponds to $\theta = \pi/2$. At this angle

$$\eta_{\max} = 2m$$

from Eq. (5.46). In much of the literature the value $2m$ is given the symbol f.

That the circle of Eq. (5.47) exists only above the ξ axis is shown from Eq. (5.46) since, for all values of θ, η remains positive.

The relations from Fig. 5.10,

$$\beta = \tan^{-1} \frac{m}{c_1}$$

and

$$a = \sqrt{c_1{}^2 + m^2} = c_1 \left(1 + \frac{1}{2} \frac{m^2}{c_1{}^2} + \cdots \right)$$

may now be used in the general lift and moment equations to derive the theoretical expressions for lift and moment. First, however, let us consider the relation between a and c_1. Assuming a large camber of 6 per cent, and letting the total chord be c,

$$\frac{f}{c} = \frac{2m}{4c_1} = 0.06 \qquad \text{and} \qquad \frac{m}{c_1} = 0.12$$

The value of $\frac{1}{2}(m^2/c_1{}^2)$, then becomes 0.0072, and $a = c_1$ within less than 1 per cent. Letting $\beta = m/c_1$ and $a = c_1$ as above, Eq. (5.27) reduces to

$$c_l = 2\pi \left(\alpha + \frac{2f}{c} \right) \tag{5.49}$$

and

$$\frac{dc_l}{d\alpha} = 2\pi$$

Since

$$\beta \cong \frac{m}{c_1} = \frac{2f}{c} \tag{5.50}$$

the angle of zero lift becomes

$$\alpha_{ZL} = -\frac{2f}{c} \tag{5.51}$$

The moment coefficient is [Eq. (5.34)]

$$c_{m\downarrow} = -\frac{\pi}{2}\beta = -\frac{\pi f}{c} \qquad (5.52)$$

The experimental results for circular-arc airfoils (usually available at extremely low Reynolds numbers only) do not agree well with theory as developed above, owing to the effects of viscosity. However, when circular-arc camber is added to a symmetrical airfoil, the agreement of theory and practice is good. For example, Eq. (5.51) yields $\alpha_{ZL} = -4.5°$ for 4 per cent camber. Experiments with airfoils with 4 per cent circular-arc camber at Reynolds number of about 2,000 000 yield nearly the same amount.

It is not to be inferred that because a circular arc is a poor airfoil the time spent analyzing it is by any means wasted. Arcs of small camber yield very nearly elliptic chord loading and are used in conjunction with no-camber airfoils in airfoil design work. More exactly, a parabolic mean line yields elliptic chord loading, but with reasonable amounts of camber (up to about 4 per cent chord) the difference between arc and parabola is exceedingly small.

As a matter of interest, the slope of the circular-arc airfoil at the three-quarter chord point is very nearly its theoretical angle of zero lift. This phenomenon is used in wind-tunnel wall corrections for streamline curvature.

Although perhaps this puts us a little ahead of ourselves, it may be noted from Fig. 5.10 that, for this particular case, both points A and B are stagnation points and that they transform to the leading and trailing edges of the airfoil. This means that in this case the flow both enters and leaves this airfoil "smoothly" at $\alpha = 0$. Leaving the trailing edge smoothly is, of course, the Kutta condition. Entering the leading edge smoothly could be called the *Theodorsen condition* since it was first noted as an important parameter by Theodorsen (Ref. 7.5). The smooth entry at the leading edge means that the pressure differential across the leading edge is then zero, and the angle of attack to which this corresponds is called the *ideal angle*. The ideal angle assumes importance later when it is shown that (1) the pressure distribution about an airfoil at any useful angle of attack α may be considered as equal to the pressure distribution at the ideal angle plus a pressure distribution corresponding to that of a flat plate at $\alpha_{\text{flat plate}} = \alpha - \alpha_I$ and (2) in the practical case the ideal angle corresponds to the middle of the "drag bucket" and to $c_{l,\text{design}}$ for a particular airfoil.

5.11. The Symmetrical Airfoil. If the center of the z circle μ is chosen so that it is a little to the right of the origin, a becomes a little larger than c_1 and the z circle transforms to a symmetrical airfoil. From Fig. 5.11, β is then zero.

Writing $a = c_1(1 + \epsilon)$ when ϵ is small and equal to m/c_1, we have as usual

$$\zeta = z + \frac{c_1{}^2}{z} = re^{i\theta} + \frac{c_1{}^2}{r}e^{-i\theta} \tag{5.53}$$

At the nose of the airfoil, $\theta = 0$, and since $r = a + m$ (from the figure) and $a = c_1 + m$, we may say

$$r = (c_1 + m) + m = c_1(1 + 2\epsilon) \tag{5.54}$$

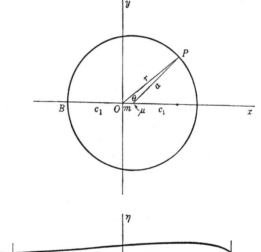

FIG. 5.11.

Substituting into Eq. (5.53), at $\theta = 0$, $e^{i\theta} = 1.0$ we have

$$\zeta = c_1(1 + 2\epsilon) + c_1(1 + 2\epsilon)^{-1}$$

which is all real, so that (clearing and expanding and neglecting higher orders of ϵ)

$$\xi \cong 2c_1(1 + 2\epsilon^2)$$

at the leading edge.

At the trailing edge of the airfoil $\theta = \pi$, $r = c_1$, and Eq. (5.53) reduces directly (since $e^{i\pi} = e^{-i\pi} = -1$) to

$$\xi = -2c_1 \tag{5.55}$$

Thus the total chord

$$c = 4c_1(1 + \epsilon^2) \tag{5.56}$$

We may sometimes make the approximation that ϵ^2 is negligible and $c = 4c_1$. For a thickness of 13 per cent, $\epsilon = 0.10$, and $c = 4.04c_1$; that is, the error in neglecting ϵ^2 would be 1 per cent in total chord.

The *equation of the airfoil* comes from the substitution of the general point P into Eqs. (5.6) and (5.7). From Fig. 5.11 and the law of cosines

$$a^2 = r^2 + m^2 - 2rm \cos \theta$$

Adding and subtracting $m^2 \cos^2 \theta$, and then assuming that m^2, which is small, is approximately equal to $m^2 \cos^2 \theta$, which is also small, we have

$$c_1 + m = r - m \cos \theta \tag{5.57}$$

or

$$r = c_1[1 + \epsilon(1 + \cos \theta)] \tag{5.58}$$

Substituting Eq. (5.58) into Eq. (5.4), we get

$$\frac{\xi}{\cos \theta} = c_1[1 + \epsilon(1 + \cos \theta)] + \frac{c_1{}^2}{c_1[1 + \epsilon(1 + \cos \theta)]}$$

Expanding the denominator of the fraction as a binomial and neglecting high orders of ϵ, we have

$$\xi = 2c_1 \cos \theta \tag{5.59}$$

which yields the approximate total chord $4c_1$. Substituting Eq. (5.58) into Eq. (5.5), we get

$$\frac{\eta}{\sin \theta} = c_1[1 + \epsilon(1 + \cos \theta)] - c_1[1 - \epsilon - \epsilon \cos \theta]$$

and

$$\eta = 2\epsilon c_1(1 + \cos \theta) \sin \theta \tag{5.60}$$

Equations (5.59) and (5.60) may be used to plot the airfoil in a manner to be more fully discussed later.

The maximum thickness of the airfoil occurs when $d\eta/d\xi = 0$. This is most simply found from

$$\frac{d\eta}{d\theta} \frac{d\theta}{d\xi} = 0 = \frac{d}{d\theta} [2\epsilon c_1(1 + \cos \theta) \sin \theta] \left[\frac{d}{d\xi} (2c_1 \cos \theta) \right]^{-1}$$

which yields

$$2 \cos^2 \theta + \cos \theta - 1 = 0$$

Solving, $\theta = \pi$, $\pi/3$. The value of π describes the minimum thickness (the trailing edge is hence a cusp), while at $\pi/3$ the maximum thickness has the value

$$\eta_{\max} = 2\epsilon c_1 \left(1 + \frac{1}{2}\right) \frac{\sqrt{3}}{2} = \frac{3\sqrt{3}}{2} \epsilon c_1 = \frac{3\sqrt{3}}{2} m$$

The total maximum thickness

$$d = 2\eta_{\max}$$

or

$$d = 3\sqrt{3}\, m$$

and the thickness ratio is

$$\frac{d}{c} = \frac{3\sqrt{3}\,\epsilon c_1}{4c_1(1 + \epsilon^2)} = \frac{3\sqrt{3}}{4} \frac{\epsilon}{(1 + \epsilon^2)} \cong \frac{3\sqrt{3}}{4}\epsilon$$

or

$$\frac{m}{c_1} = 0.77\,\frac{d}{c} \tag{5.61}$$

Equation (5.61) can be used to determine m for the actual construction of an airfoil with a selected value of d/c.

The *point of maximum thickness* occurs when $\theta = 60°$ and

$$\xi_{\text{max thickness}} \cong 2c_1 \cos\theta = c_1$$

Then

$$\frac{c_1}{c} \cong 0.25 \tag{5.62}$$

That is, the maximum thickness of all Joukowski airfoils of reasonable thickness is quite close to the quarter chord.

The *lift coefficient* may as usual be developed from Eq. (5.27), yielding in this case

$$L = \frac{\rho}{2} V^2 \cdot 4c_1(1 + \epsilon)\, 2\pi\alpha \tag{5.63}$$

and

$$c_l = \frac{2\pi(1 + \epsilon)}{(1 + \epsilon^2)}\,\alpha \tag{5.64}$$

The *slope of the lift curve* is then

$$\frac{dc_l}{d\alpha} = \frac{2\pi(1 + \epsilon)}{(1 + \epsilon^2)} \tag{5.65}$$

We see that *the slope of the lift curve increases with airfoil thickness*, being 2π per radian for a very thin airfoil and $1.11\,(2\pi)$ for an airfoil

15 per cent thick. In practice, the increase of lift curve slope with thickness is realized up to a thickness-chord ratio of about 20 per cent, but the actual values are slightly less than those given by Eq. (5.65). Theory and experiment are compared in Fig. 5.12, where it will be noted

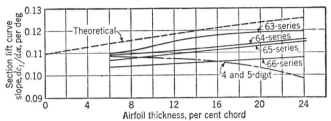

Fig. 5.12. Comparison of experimental and theoretical values for the slope of the life curve for two-dimensional airfoils. The theoretical values are those of Eq. (5.65) for symmetrical Joukowski airfoils. The experimental values are from Ref. 7.6. (*Reproduced by permission from "Principles of Aerodynamics" by James H. Dwinnell, McGraw-Hill Book Company, Inc., New York.*)

that the lift curve slope decreases as the trailing-edge angle increases.

The moment coefficient, from Eq. (5.34), is

$$c_{m\frac{1}{4}} = -\frac{\pi}{2}\beta = 0 \qquad (5.66)$$

which puts the center of pressure and the aerodynamic center at the quarter chord to the accuracy that $c = 4c_1$.

If, however, we hold additional terms of the above series, we arrive at a more accurate value as follows:

From Eq. (5.30)

$$M_\mu = \frac{\rho}{2}V^2 \cdot 4\pi c_1{}^2 \cdot \sin 2\alpha \qquad (5.67)$$

where μ is at a distance of $2c_1 + m = (2 + \epsilon)c_1$ from the airfoil trailing edge (see Fig. 5.13).

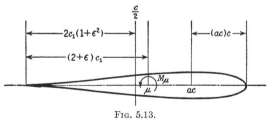

Fig. 5.13.

Letting a.c. = distance in chord lengths from the airfoil leading edge to the aerodynamic center, the moment about the aerodynamic center will be

$$M_{ac} = M_\mu - L[c - (2 + \epsilon)c_1 - (\text{a.c.})\,c]$$

Substituting from Eq. (5.64) and (5.67), and dividing by $\frac{\rho}{2} V^2[4c_1(1 + \epsilon^2)]^2$ to reduce to coefficient form, we get

$$c_{mac} = \frac{2\pi\alpha}{4(1 + \epsilon^2)^2} - \frac{2\pi\alpha(1 + \epsilon)}{4(1 + \epsilon^2)^2} [4(1 + \epsilon^2) - (2 + \epsilon) - 4(\text{a.c.})(1 + \epsilon^2)]$$

(5.68)

Applying the condition that $dc_{mac}/d\alpha = 0$, Eq. (5.68) reduces to

$$\text{a.c.} = 1 - \frac{1}{4(1 + \epsilon^2)} \left(2 + \epsilon + \frac{1}{1 + \epsilon}\right)$$

(5.69)

or, keeping terms to the order of ϵ^2,

$$\text{a.c.} = \frac{1}{4} + \frac{\epsilon^2}{2}$$

(5.70)

Thus for a 9 per cent thick airfoil the aerodynamic center is theoretically at 25.2 per cent chord or, for a 15 per cent thick airfoil, at the 25.6 per cent point. In practice (Fig. 5.14), aerodynamic-center points

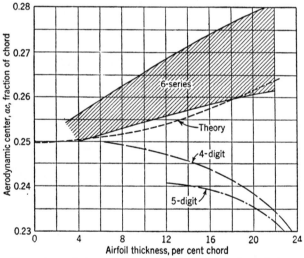

FIG. 5.14. Comparison of theoretical and experimental locations of the aerodynamic center. Theoretical values are for a two-dimensional Joukowski symmetrical airfoil, experimental values from Ref. 7.6. (*Reproduced with minor modifications by permission from "Principles of Aerodynamics" by James H. Dwinnell, McGraw-Hill Book Company, Inc., New York.*)

are found both ahead and behind the theoretical locations. No complete explanation has been advanced for this phenomenon.

Substituting Eq. (5.69) into Eq. (5.68), we see that

$$c_{mac} = 0$$

for a symmetrical Joukowski airfoil, and indeed this conclusion holds for all symmetrical sections.

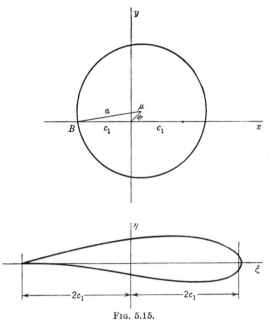

Fig. 5.15.

5.12. The Cambered Airfoil. When the center of the z circle is to the right of and above the origin, the Joukowski transformation yields a cambered airfoil. From Fig. 5.16 it is seen that (approximately)

$$|a| = |c_1| + |\mu\mu'|$$

Now since the angle β is small, $\mu\mu'$ may be considered approximately equal to the m of the symmetrical section: that is, $\mu\mu'$ becomes the quantity that determines the thickness of the airfoil. Hence

$$\frac{d}{c} = 1.30 \frac{\mu\mu'}{c_1} \quad \text{or} \quad \epsilon = 0.77 \frac{d}{c}$$

and

$$a = c_1(1 + \epsilon)$$

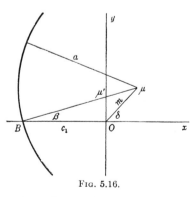

Fig. 5.16.

Similarly, $O\mu'$ may be assumed approximately equal to the m part of the circular-arc airfoil, making the mean-line camber $f = 2O\mu'$. Then by Eq. (5.49)

$$\beta = \frac{2f}{c}$$

The general relations may be used again to find the lift and moment coefficients.

$$L = \frac{\rho}{2}V^2 \cdot \frac{4c_1(1+\epsilon)}{1+\epsilon^2} \cdot 2\pi \sin\left(\alpha + \frac{2f}{c}\right)$$

and

$$c_l = 2\pi \frac{1+\epsilon}{1+\epsilon^2} \sin\left(\alpha + \frac{2f}{c}\right) \tag{5.71}$$

$$\frac{dc_l}{d\alpha} = 2\pi \frac{1+\epsilon}{1+\epsilon^2} \tag{5.72}$$

$$\alpha_{ZL} = -\frac{2f}{c} \tag{5.73}$$

$$c_{m\frac{1}{4}} = -\frac{\pi f}{c} \tag{5.74}$$

The equation of the airfoil becomes somewhat involved for the cambered airfoil and will not be presented. However, if it is desired to plot one, a graphical method of good accuracy will be found in Chap. 6.

Equations (5.73) and (5.74) give good agreement with the actual performance of the Joukowski. For example, with 4 per cent camber (at the 50 per cent chord point) the calculated values of α_{ZL} and $c_{m\frac{1}{4}}$ are $-4.5°$ and -0.125. A nearly equivalent airfoil (the NACA 4512) developed $-4.2°$ and -0.106 at an $RN_e = 8{,}000{,}000$. The small discrepancy is easily attributed to the difference in mean camber line (a circular arc for the Joukowski and a parabola for the NACA 4512) and to the boundary layer.

PROBLEMS

5.1. Does the total length of a closed curve necessarily remain constant if the curve is conformally transformed?

5.2. What is the Kutta condition? What important problem does it solve?

5.3. Do we actually find the potential function of the flow about the Joukowski airfoil? Why?

5.4. When the circle to be transformed has its center at me^{i0}, a symmetrical airfoil results. Explain why the portions of the airfoil that correspond to the first and second quadrants of the circle are dissimilar.

5.5. Give three important conclusions about moment coefficients and aerodynamic centers.

5.6. Give the theoretical values of c_l and $c_{m\frac{1}{4}}$ for a flat plate at an angle of attack of 4°.

5.7. According to the Joukowski theory, what $c_{m\frac{1}{4}}$ should we expect for a Clark Y airfoil if its angle of zero lift is $-5°$?

5.8. Calculate α_{ZL}, $c_{m\frac{1}{4}}$, for a circular-arc profile with 3 per cent camber, and find its center of pressure at $c_l = 0.4$.

5.9. Calculate the slope of the lift curve for a symmetrical airfoil (Joukowski) 14 per cent thick. Compare your results with the theoretical value for a flat plate.

5.10. If 3 per cent camber is added to the airfoil of Prob. 5.9, find c_l at $\alpha = 4°$.

5.11. From your knowledge of a Joukowski airfoil and the NACA four-digit nomenclature, what NACA four-digit airfoil best approximates a 15 per cent thick Joukowski airfoil with 4 per cent camber?

5.12. Calculate α_{ZL} and $c_{m\frac{1}{4}}$ for a circular-arc profile with 4 per cent camber, and find its center of pressure at $c_l = 0.4$.

5.13. Plot an NACA 4512 and a 12 per cent thick Joukowski with 4 per cent camber on the same axes, and note differences in thickness distribution and camber.

5.14. Plot a symmetrical Joukowski airfoil 11 per cent thick with a 10-ft chord. Assume $c = 4c_1$.

REFERENCES

5.1. V. L. Streeter, "Fluid Dynamics," pp. 93–100, McGraw-Hill Book Company, Inc., New York, 1948.

5.2. W. F. Durand, "Aerodynamic Theory," Vol. I, pp. 89–104, Verlag Julius Springer, Berlin, 1934.

5.3. Leonard Bairstow, "Applied Aerodynamics," p. 325, Longmans, Green & Co., Inc., New York.

5.4. G. Eiffel, "Resistance of the Air and Aviation," Houghton Mifflin Company, Boston.

5.5. H. Reese Ivey, E. Bernard Klunker, and Edwin N. Bowen, A Method for Determining the Aerodynamic Characteristics of Two and Three Dimensional Shapes at Hypersonic Speeds, *TN* 1613, July, 1948.

CHAPTER 6

AIRFOIL CONSTRUCTION AND PRESSURE DISTRIBUTION

6.1. Trefftz Graphical Construction. Although the equations for the flat-plate, circular-arc, and symmetrical Joukowski airfoils are simple and direct, those of the cambered airfoil are complicated and may be by-passed only because an ingenious graphical method exists which may be used to find both the airfoil *and* its pressure distribution. The

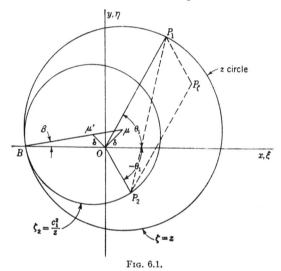

FIG. 6.1.

graphical method, due to Trefftz (Ref. 6.1) has the following development (see Fig. 6.1):

Consider a z circle with its center at $me^{i\delta}$, and write the Joukowski transformation as

$$\zeta = \zeta_1 + \zeta_2 = z + \frac{c_1^2}{z} \tag{6.1}$$

so that

$$\zeta_1 = z \quad \text{and} \quad \zeta_2 = \frac{c_1^2}{z} \tag{6.2}$$

Each point on the z circle now is transformed twice, becoming the points $P_{\zeta 1}$ and $P_{\zeta 2}$, whose vector sum is the desired point P_ζ. It is of interest that both transformations yield circles.

100

The equation of the z circle is obviously

$$(x - m \cos \delta)^2 + (y - m \sin \delta)^2 = a^2 \tag{6.3}$$

and the first transformation makes it an identical circle in the ζ_1 plane described by

$$(\xi_1 - m \cos \delta)^2 + (\eta_1 - m \sin \delta)^2 = a^2 \tag{6.4}$$

The second transformation may be applied best to Eq. (6.3) in another form. Expanding Eq. (6.3), we have

$$x^2 + y^2 - 2mx \cos \delta - 2my \sin \delta + m^2 - a^2 = 0$$

Substituting from Eq. (2.6), this becomes

$$z\bar{z} - m(z + \bar{z}) \cos \delta + im(z - \bar{z}) \sin \delta + m^2 - a^2 = 0$$

which, by the second transformation, is

$$\frac{c_1^4}{\zeta_2 \bar{\zeta}_2} - mc_1^2 \left(\frac{\bar{\zeta}_2 + \zeta_2}{\zeta_2 \bar{\zeta}_2} \right) \cos \delta + imc_1^2 \left(\frac{\bar{\zeta}_2 - \zeta_2}{\zeta_2 \bar{\zeta}_2} \right) \sin \delta + (m^2 - a^2) = 0$$

Dividing by $(m^2 - a^2)$ and multiplying by $\zeta_2 \bar{\zeta}_2$, we have

$$\frac{c_1^4}{m^2 - a^2} - \frac{mc_1^2}{m^2 - a^2} (\bar{\zeta}_2 + \zeta_2) \cos \delta + i \frac{mc_1^2}{m^2 - a^2} (\bar{\zeta}_2 - \zeta_2) \sin \delta + \zeta_2 \bar{\zeta}_2 = 0$$

Again applying Eq. (2.6), this time worked out for $\zeta_2 = \xi_2 + i\eta_2$, we get

$$\zeta_2 - \bar{\zeta}_2 = 2i\eta_2, \qquad \zeta_2 + \bar{\zeta}_2 = 2\xi_2, \qquad \zeta_2 \bar{\zeta}_2 = \xi_2^2 + \eta_2^2$$

and

$$\frac{c_1^4}{m^2 - a^2} - \frac{2mc_1^4 \cos \delta}{m^2 - a^2} \xi_2 + \frac{2mc_1^2 \sin \delta}{m^2 - a^2} \eta_2 + \xi_2^2 + \eta_2^2 = 0$$

or

$$\left(\xi_2 + \frac{mc_1^2 \cos \delta}{a^2 - m^2} \right)^2 + \left(\eta_2 - \frac{mc_1^2 \sin \delta}{a^2 - m^2} \right)^2 = \frac{c_1^4 a^2}{a^2 - m^2} \tag{6.5}$$

This is a circle with its center μ' at

$$-\frac{mc_1^2 \cos \delta}{a^2 - m^2}, \qquad \frac{mc_1^2 \sin \delta}{a^2 - m^2}, \tag{6.6}$$

and radius

$$R' = \frac{ac_1^2}{a^2 - m^2} \tag{6.7}$$

Consider the position vector of μ', $O\mu'$. From Eq. (6.6) the modulus is $mc_1{}^2/(a^2 - m^2)$, and since

$$\cos(\pi - \delta) = -\cos\delta \qquad (6.8)$$
$$\sin(\pi - \delta) = \sin\delta$$

the argument is $(\pi - \delta)$.

Thus the line $O\mu'$ lies along the reflection of the line $O\mu$ about the η axis. The location of μ' with respect to the line $B\mu$ drawn to μ from B

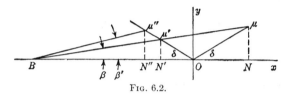

Fig. 6.2.

may be determined by assuming that μ' is elsewhere than at the intersection of $B\mu$ and $O\mu'$, say at μ''. Then (Fig. 6.2)

$$\tan\beta' = \frac{N''\mu''}{B\mu''} = \frac{mc_1{}^2\sin\delta}{a^2 - m^2} \div \left(c_1 - \frac{mc_1{}^2\cos\delta}{a^2 - m^2}\right)$$
$$= \frac{mc_1{}^2\sin\delta}{c_1(a^2 - m^2) - mc_1\cos\delta} \qquad (6.9)$$

But in triangle $BO\mu$,

$$B\mu = a \qquad \text{and} \qquad a^2 = c_1{}^2 + m^2 + 2mc_1\cos\delta$$

from which

$$c_1(a^2 - m^2) = c_1{}^2(c_1 + 2m\cos\delta)$$

So from Eq. (6.9)

$$\tan\beta' = \frac{m\sin\delta}{c_1 + m\cos\delta} \qquad (6.10)$$

But from triangle $B\mu N$

$$\tan\beta = \frac{N\mu}{BO + ON} = \frac{m\sin\delta}{c_1 + m\cos\delta} \qquad (6.11)$$

so that $\beta' = \beta$ and μ' must be coincident with μ''. Thus the centers of the ζ_1 and ζ_2 circles lie on the same line and have the rear stagnation point B in common, making the circles internally tangent. The larger ζ_1 circle is usually called the *major* circle and the ζ_2 circle the *minor* circle. One point of interest remains: the relation $\zeta_1 = z$ determines that a point P_1 on the major circle remains unchanged by the transformation, but the relation

$$\zeta_2 = \frac{c_1{}^2}{z} = \frac{c_1{}^2}{r}e^{-i\theta} \tag{6.12}$$

places the transformed point P_z on the minor circle at $-\theta$, and hence the transformed final point P_ζ is the resultant of the vectors OP_1 and OP_2 as shown in Fig. 6.3.

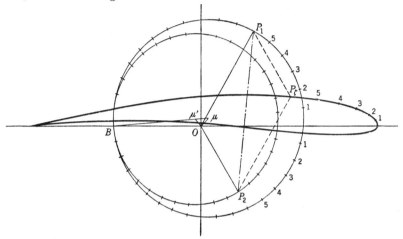

FIG. 6.3. Construction of a Joukowski airfoil with 4-ft chord, 4 per cent camber, 10 per cent thick, by the Trefftz method.

The particular values of m and β have been discussed in Chap. 5 and may be determined for selected values of camber, chord, and thickness. A general example is shown below:

Example 6.1. Construct a Joukowski airfoil 10 per cent thick with 4 per cent camber and a 4-ft chord.

1. Draw ξ, η axes. From Eq. (5.61), $m = 0.77(d/c)c_1 = 0.077c_1$. From Eq. (5.56), $c = 4 = 4c_1[1 + (m/c_1)^2] = 4c_1(1 + 0.077^2)$, and $c_1 = 0.995$. Then locate B at c_1.

2. At B construct $\beta = 2f/c = 2(0.04)(4)/4 = 0.08$ radian $= 4.58°$.

3. Calculate the radius of major circle a.

$$a = c_1\left(1 + 0.77\frac{d}{c}\right) = 0.995 + \frac{(0.77)(0.4)(0.995)}{4}$$
$$= 1.072 \text{ ft}$$

4. Lay off $B\mu = a = 1.072$ ft. Draw $O\mu$, which will be at angle δ to the ξ axis.

5. Construct angle $BO\mu' = $ angle δ. μ' is the center of the minor circle.

6. Construct major and minor circles through point B.

7. Draw OP_1 at any angle θ and OP_2 at $-\theta$.

8. Add vectors OP_1 and OP_2 to obtain point P_ζ on the airfoil.

6.2. Pressure Distributions. The static pressure distribution over
the surface of an airfoil yields directly the lift and pressure drag. Ordinarily the pressures are read or calculated, atmospheric pressure is subtracted, and the remainder divided by the dynamic pressure in order to

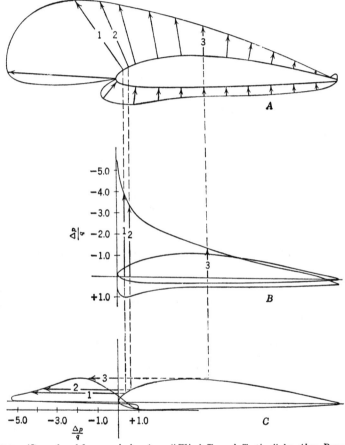

FIG. 6.4. *(Reproduced by permission from "Wind Tunnel Testing" by Alan Pope, John Wiley & Sons, Inc., New York.)*

form a nondimensional coefficient. The coefficients are then plotted perpendicularly to the chord at the appropriate stations for a normal force
plot or parallel to the chord at the appropriate thickness stations for a
chord force plot.* Integration of the pressure plots will yield the normal

* The plots of pressure coefficients against chord or thickness stations do not change
with changes in dynamic pressure and hence are more useful and general than plots
of pressures against chord stations.

and chord force coefficients. This process is illustrated in Fig. 6.4, where the actual (A), normal (B), and chordwise (C) pressure plots are compared. Several of the pressure force vectors are labeled so that their relative positions may be followed in the three plots.

The explanation of the validity of plotting the pressure forces normal to the two axes instead of taking their horizontal and vertical components is as follows:

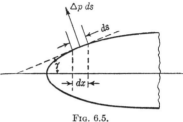

Consider (Fig. 6.5) a small element of surface ds which is subjected to a static pressure increment $\Delta p = p - p_0$ acting normal to it. The total force on the element is $\Delta p\, ds$, directed along the normal to the surface, and the component of this force normal to the chord line is

FIG. 6.5.

$\Delta p\, ds \cos \gamma$. But $ds \cos \gamma = dx$, where dx is a short length of the chord c, so that the total force normal to the wing chord is

$$N = \int_0^c \Delta p\, dx$$

By definition

$$N = \frac{\rho}{2} SV^2 c_N$$

and hence

$$c_N = \frac{N}{qc} = \int_0^{1.0} \frac{\Delta p}{q} \frac{dx}{c} \tag{6.13}$$

It follows that the curve of pressures plotted in units of dynamic pressure against the fractional chord may be integrated to get the normal force coefficient.

Let us now consider a low-angle pressure distribution (see Fig. 6.6) and outline the information it yields in order better to understand the pressure-distribution work to follow:

First of all consider the magnitude of the maximum pressures, both positive and negative. In general there will be a stagnation point near the airfoil leading edge where the velocity is zero and $\Delta p/q = +1.0$. Two exceptions to the general case exist. The first is the case of swept-back wings where crossflow is such that full stagnation does not realize, and the second concerns testing at speeds where compressibility is not negligible. In the latter case pressure coefficients based on $q = \frac{1}{2}\rho V^2$ may easily exceed $+1.0$.

Minimum pressure usually occurs near the leading edge when the airfoil is at high angle of attack. At that time pressure coefficients of

−5.0 to −8.0 can develop, indicating very high local velocities. It is usually advantageous to avoid high negative pressures, especially at cruising angle of attack, as the resultant velocity for a fast airplane might be far enough above sonic speed to cause excessive drag.

The pressures, either negative or positive, yield the direct loading that the skin must withstand.

The sign and magnitude of the rate of pressure variation along the chord also lead to important factors. As long as the pressure is falling (see ABC and $AB'C'$ in Fig. 6.6), the boundary-layer conditions are most satisfactory, and it is possible that in turbulence-free air with a smooth, vibration-free airfoil a laminar boundary layer will exist until the minimum pressure points (C and C') are reached. For a given amount of surface roughness, vibration, or turbulence, laminar boundary-layer flow is more apt to exist where the gradient is steep (AB and AB') rather than gradual.

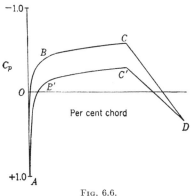

FIG. 6.6.

The positive (rising) pressure gradient acts in an opposite manner and seeks to restrain the progress of the air over the wing. When the gradient becomes strong enough, actual flow separation is caused and the smooth air leaves the proximity of the wing with large losses in lift and increases in drag. The rapid rise in pressure close to the trailing edge usually is severe enough to cause separation and accounts for the difference between the actual and theoretical pressure distribution. This particular separation does not affect the lift as would separation farther forward.

The remaining factors of importance concern the shape of the normal-force pressure-distribution plot. Since the moment of area about the leading edge divided by the area yields the center of pressure, it is seen that an increase of area near the trailing edge will increase the airfoil pitching moment. Such an increase (caused by camber as seen by comparing Figs. 6.8 and 6.13) is usually undesirable since in use it would lead to aircraft with excessively large tails. In the chapters to follow the student will doubtless be impressed by the control possible over the factors described above.

6.3. Pressure Distribution over a Symmetrical Joukowski Airfoil. The value of the velocity in the ζ plane has already been expressed in Eq. (5.9).

However, before actual values of q_ζ can be calculated, additional information on the interpretation of θ is in order. To begin with, the velocity at a point on the surface of a circular cylinder with circulation is [Eq. (4.17)]

$$q_z = q_{\theta'} = 2V \sin \theta' + \frac{\Gamma}{2\pi a}$$

where $\Gamma = 4\pi a V \sin(\alpha + \beta)$
  a = radius of cylinder
  θ' = angle between the direction of the free stream and the radius of the circle to point P

From Fig. 6.7

$$\theta' = \tan^{-1} \frac{r \sin \theta}{r \cos \theta - m} + \alpha \qquad (6.14)$$

where α = angle between $\theta = 0$ and the free stream.

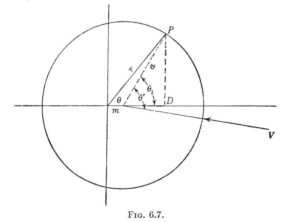

Fig. 6.7.

Using Eqs. (5.9) and (5.10), we may find the velocity q_ζ in the ζ plane. The local pressure coefficient comes directly from Bernoulli's equation, written

$$C_p = 1 - \left(\frac{q_\zeta}{V}\right)^2 \qquad (6.15)$$

The actual mechanics of completing a pressure-distribution calculation leads to a table some 36 columns in length; the first 16 columns are for the transformation and need be figured only once for a given airfoil, while the remainder are entirely different for each angle of attack. A total of about 16 man-hours is required for the transformation and the pressure distribution at one angle of attack, including plotting the chord and normal-force pressure distributions.

Example 6.2. Plot the pressure distribution about an 11 per cent thick symmetrical Joukowski airfoil with a 10-ft chord at 4° angle of attack. Assume $V = 100$ ft per sec and $c = 4c_1$.

a. From Eq. (5.61)

$$m = 0.77 \frac{d}{c} c_1 = (0.77)(0.11)(2.5) = 0.212$$

$$a = c_1 + m = 2.50 + 0.212 = 2.712$$

b. To save space, the pressure coefficient will be calculated for one θ only. The angle θ varies from 0 to 180° for the upper surface and 0 to $-180°$ for the lower surface. Fifteen-degree increments should suffice for the bulk of the cylinder, as long as much smaller intervals are taken near the points corresponding to the leading and trailing edges. For the angle $\theta = 30°$ we have the following:

1. $\theta = 30°$
2. $\sin \theta = 0.5000$
3. $\cos \theta = 0.866$
4. $\xi = 2c_1 \cos \theta = 4.33$
5. $\xi \ \%$ chord $= \dfrac{5 - 4.33}{10} = 6.67$
6. $(1 + \cos \theta) = 1.866$
7. $2\epsilon c_1 (1 + \cos \theta) = 0.792$
8. $2\epsilon c_1 (1 + \cos \theta) \sin \theta = 0.396$
9. $\eta, \ \%$ chord $= 3.96$
10. $m \cos \theta = 0.184$
11. $r = c_1 + m + m \cos \theta = 2.896$ [Eq. (8.47)]
12. $r \cos \theta = 2.505$
13. $r \sin \theta = 1.448$
14. $r \cos \theta - m = 2.293$
15. $\dfrac{r \sin \theta}{r \cos \theta - m} = 0.6306$
16. $\tan^{-1} \dfrac{r \sin \theta}{r \cos \theta - m} = 32.25° = \theta''$
17. $\theta' = \theta_1 + \alpha = 36.25°$
18. $\sin \theta' = 0.591$
19. $2V \sin \theta' = 118.4$
20. $q_s = 2V \sin \theta' + \dfrac{\Gamma}{2\pi a} = 132.36$
21. $r^2 = 8.39$
22. $\dfrac{c_1{}^2}{r^2} = 0.7453$
23. $2\theta = 60$
24. $\sin 2\theta = 0.866$
25. $\cos 2\theta = 0.500$
26. $\dfrac{c_1{}^2}{r^2} \cos 2\theta = 0.373$
27. $\dfrac{c_1{}^2}{r^2} \sin 2\theta = 0.6454$
28. $1 - \dfrac{c_1{}^2}{r^2} \cos 2\theta = 0.6273$
29. $\left(1 - \dfrac{c_1{}^2}{r^2} \cos 2\theta\right)^2 = 0.3935$
30. $\left(\dfrac{c_1{}^2}{r^2} \sin 2\theta\right)^2 = 0.417$
31. Item 29 + item 30 = 0.810
32. $\left|\dfrac{d\zeta}{dz}\right| = \sqrt{\text{item } 31} = 0.900$
33. $q_\zeta = \dfrac{\text{item } 20}{\text{item } 32} = 147.1$ ft per sec
34. $\dfrac{q_\zeta}{V} = 1.471$
35. $\left(\dfrac{q_\zeta}{V}\right)^2 = 2.163$
36. $C_p = -1.163$

It will be noted that in this analytical work-up the approximation $c = 4c_1$ was used, although in Ex. 6.1 the more exact relation was employed. The reason is that the Trefftz construction is exact, while Eqs. (5.59) and (5.60) neglect the higher order ϵ^2 terms. If it is desired

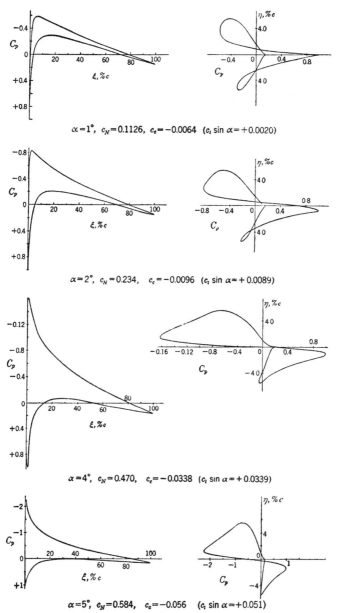

$\alpha = 1°$, $c_N = 0.1126$, $c_c = -0.0064$ ($c_i \sin \alpha = +0.0020$)

$\alpha = 2°$, $c_N = 0.234$, $c_c = -0.0096$ ($c_i \sin \alpha = +0.0089$)

$\alpha = 4°$, $c_N = 0.470$, $c_c = -0.0338$ ($c_i \sin \alpha = +0.0339$)

$\alpha = 5°$, $c_N = 0.584$, $c_c = -0.056$ ($c_i \sin \alpha = +0.051$)

Fig. 6.8. Normal and chordwise pressure distribution for an 11 per cent thick symmetrical Joukowski airfoil at several angles of attack. Values given are actual integration values and compare closely with the theoretical values using Eq. (5.65).

to include ϵ^2 terms in ξ, Eq. (5.60) should be expanded to include them in η.

A plot of Ex. 6.2 including the remainder of the chord points and the pressure distributions for several other angles are shown in Fig. 6.8. From the resolution of c_V and c_l we should expect $c_C \cong -c_N \sin \alpha$.

6.4. Pressure-distribution Discussion. Using the method of Sect. 6.3, the pressure distributions for the simple flat-plate and arc airfoils of Sects. 5.10 and 5.11 have been calculated and are presented in Figs. 6.9 and 6.10. The item of greatest interest is that, when the $\alpha = 0°$ distribution of the circular-arc airfoil is subtracted from that at $\alpha = 3°$,

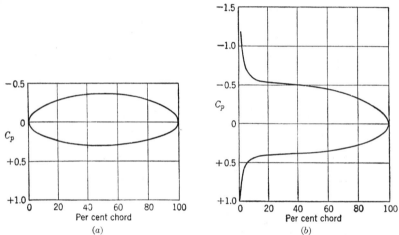

Fig. 6.9. Normal-force pressure distribution for a circular-arc airfoil at $\alpha = 0$ (a) and $\alpha = 3°$ (b).

the remainder very closely approximates the normal-force pressure distribution over a flat plate at $\alpha = 3°$ (see Fig. 6.11). Hence, if the pressure distribution is known for some "ideal" angle of attack (to be defined and discussed in Chap. 7), that at some new angle of attack may be found by adding the flat-plate distribution for the new angle. The practical implications are quite obvious.

Another practical application of the "additional-" pressure-distribution theory is found in the study of control-surface hinge moments. One of the important criteria of control-surface design is the rate of change of hinge moment with angle of attack $C_{H\alpha}$, and so-called "balance area" ahead of a control-surface hinge line is usually provided to reduce the moment. Drawing an airfoil on the same axis as the total pressure coefficient due to angle of attack, we see immediately that, the farther forward the balance is located, the greater will be the reduction of hinge moment since the balance receives a greater pressure differential across it. In other words, the $C_{H\alpha}$ of a 30 per cent flap with a 15 per cent balance will be less than that of a 20 per cent flap with a 15 per cent balance. Since the effectiveness of a balance is a function of the product of the

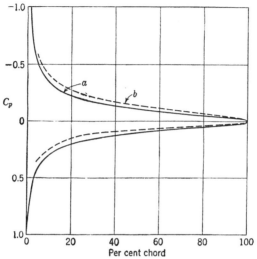

FIG. 6.10.   The pressure distribution over a flat plate at 3° α (solid line *a*) compared with the *difference* between the pressure distribution over a circular-arc airfoil with 4 per cent camber at 0 and at 3° α (broken line *b*).

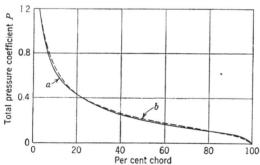

FIG. 6.11.   The total pressure over a flat plate at 3° α (solid line *a*) compared with the *difference* between the total pressure over a circular-arc airfoil with 4 per cent camber at 0 and at 3° α (broken line *b*).

area times the pressure differential, we expect (and get) relatively small changes in $C_{H\alpha}$ with changes in balance profile.

Considering the hypothesis that the flat-plate pressure distribution is added to the ideal distribution by increased angle of attack, those interested in the reduction of drag by extension of the laminar boundary layer will note that the negative slope of the pressure-distribution curve over the forward part of the airfoil at zero lift must be greater than the positive slope shown by the flat plate if a favorable gradient is to continue for the maximum angle of attack range. Further, since the slope of the forward part of the pressure-distribution curve at the ideal angle of attack is increased by airfoil thickness, we may rightly infer that the angle-of-attack range through which we can expect an extended laminar boundary layer is increased by airfoil thickness.

6.5. Pressure Distribution by the Trefftz Method. The expression

$$|q_\zeta| = \frac{|q_z|}{\left|\dfrac{d\zeta}{dz}\right|}$$

is more simply solved graphically by an extension of the Trefftz method. The method also applies to cambered airfoils as well as symmetrical ones.

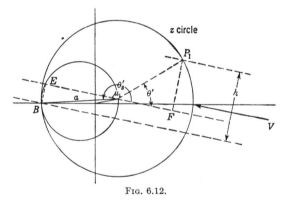

FIG. 6.12.

Let us first find the value of q_z. Consider Eq. (4.17),

$$q_z = 2V \sin \theta' + \frac{\Gamma}{2\pi a} \tag{6.16}$$

When the point under consideration is the stagnation point, $q_z = 0$ and θ' becomes, say, θ_s'. Then

$$\frac{\Gamma}{2\pi a} = -2V \sin \theta_s'$$

Substituting into Eq. (6.16),

$$q_z = 2V(\sin \theta' - \sin \theta_s') \tag{6.17}$$

Consider Fig. 6.12, upon which a Trefftz major circle has been constructed and dotted lines parallel to the relative wind through the center of the major circle and the stagnation point have been added. From the figure,

$$\sin \theta_s' = - \sin (\theta_s' - 180) = - \frac{EB}{\mu_1 B}$$

and

$$\sin \theta' = \frac{FP}{\mu_1 P} = \frac{FP}{\mu_1 B}$$

And from Eq. (6.17)

$$q_z = 2V \frac{FP + EB}{\mu_1 B} = 2V \frac{h}{a} \tag{6.18}$$

The values of h and a may be simply measured and q_z determined for all selected values of θ' or θ. The value of $\frac{d\zeta}{dz}$ may be found by writing

$$\frac{d\zeta}{dz} = 1 - \frac{c_1^2}{z^2} = \frac{1}{z}\left(z - \frac{c_1^2}{z}\right) \tag{6.19}$$

Since we have from the Trefftz construction that $\zeta_1 = z$ and $\zeta_2 = c_1^2/z$, Eq. (6.19) becomes

$$\frac{d\zeta}{dz} = \frac{\zeta_1 - \zeta_2}{\zeta_1} \tag{6.20}$$

The value of ζ_1 is simply (see Fig. 6.1) OP_1 and that of $\zeta_1 - \zeta_2$ simply P_1P_2, so that

$$\left|\frac{d\zeta}{dz}\right| = \left|\frac{P_1P_2}{OP_1}\right| \tag{6.21}$$

Finally,

$$q_\zeta = \frac{|q_z|}{\left|\frac{d\zeta}{dz}\right|} = 2V \left|\frac{h}{a}\right| \left|\frac{OP_1}{P_1P_2}\right| \tag{6.22}$$

Equation (6.22) is very instructive. First of all h and P_1P_2 are both zero at the trailing edge, making the velocity at that point indeterminate, as previously discussed in Sect. 5.4. When $\alpha = 0$, $h = 0$ at the point corresponding to the leading edge and q_ζ is then zero. This condition corresponds to the Theodorsen condition (see Chap. 7), and we may hence infer that the ideal angle of a Joukowski airfoil is 0°.

Using Eq. (6.22), the velocity at each particular point in the ζ plane may be found as a function of its original θ. However, before the pressure plots can be made, two additional items are required: (1) the pressure coefficient corresponding to the velocity q_ζ and (2) the values of ξ and η in per cent chord corresponding to the selected θ.

The solution of (1) is simply the velocity-pressure relation of Eq. (3.10) and that for (2) the location of ζ_1 corresponding to θ_1. The vector P_1P_2 is shown in Fig. 6.1 as is OP_1. The normal-force pressure distribution for the airfoil of Fig. 6.3 at $\alpha = 5°$ is shown in Fig. 6.13, while a com-

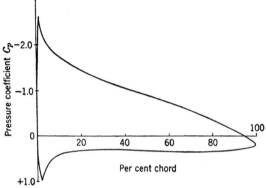

FIG. 6.13. Normal-force pressure distribution over an 11 per cent thick Joukowski airfoil with 4 per cent camber. $\alpha = 4°$. Integrated $c_N = 1.26$, which closely checks the value by Eq. (5.71).

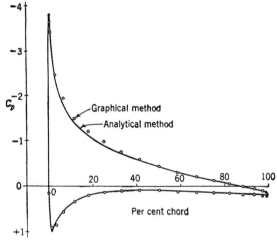

FIG. 6.14. Comparison of normal-force pressure distribution over an 11 per cent thick symmetrical Joukowski airfoil at 7° angle of attack by analytical and graphical methods.

parison of the analytical and the graphical method of finding pressure distribution is made in Fig. 6.14.

6.6. Kármán-Trefftz Airfoils. We have seen how the relatively simple Joukowski transformation can be made to yield a family of airfoils and, directly, their pressure distributions. But the Joukowski airfoils with the maximum thickness near the quarter chord do not show up well under actual test; further, the cusp trailing edge is both difficult to construct and poor aerodynamically if control surfaces are to be used. We might say that another transformation is needed that will yield more practical airfoils, but in actuality the problem is more that of empirically determining desirable pressure distributions and then the airfoil shapes that will yield them. For completeness we shall continue on the "direct" problem.

Several transformations exist that will yield better airfoils than the Joukowski, and of these the von Mises and Kármán-Trefftz are well known. We shall consider the Kármán-Trefftz procedure since it follows directly from the work we have done so far.

Consider the Joukowski transformation

$$\zeta = z + \frac{c_1^2}{z} \tag{6.23}$$

and add $2c_1$ to each side, getting

$$\zeta + 2c_1 = \frac{z^2 + 2c_1 z + c_1^2}{z} \tag{6.24}$$

Similarly, subtracting $2c_1$ from each side yields

$$\zeta - 2c_1 = \frac{z^2 - 2c_1 z + c_1^2}{z} \tag{6.25}$$

Dividing Eq. (6.24) by Eq. (6.25), we have

$$\frac{\zeta + 2c_1}{\zeta - 2c_1} = \left(\frac{z + c_1}{z - c_1}\right)^2$$

which by inspection is a special case of

$$\frac{\zeta + nc_1}{\zeta - nc_1} = \left(\frac{z + c_1}{z - c_1}\right)^n \tag{6.26}$$

Taking the logarithm of Eq. (6.26), we have

$$\ln(\zeta + nc_1) - \ln(\zeta - nc_1) = n \ln(z + c_1) - n \ln(z - c_1) \tag{6.27}$$

and differentiating

$$d\zeta \left(\frac{1}{\zeta + nc_1} - \frac{1}{\zeta - nc_1}\right) = n \left(\frac{1}{z + c_1} - \frac{1}{z - c_1}\right) dz$$

which simplifies to

$$\frac{d\zeta}{dz} = \frac{\zeta^2 - n^2 c_1^2}{z^2 - c_1^2} \tag{6.28}$$

When n is taken slightly less than 2.0, Eqs. (6.26) and (6.28) may be used in a manner paralleling the Joukowski discussion to yield a family of airfoils whose maximum thickness may be arbitrarily set at 25 to 50 per cent chord, the trailing-edge angles increasing as the point of maximum thickness is moved back. For each maximum-thickness station and amount there is only one Kármán-Trefftz profile, so that the family while broader than the Joukowski group is still by no means arbitrary.

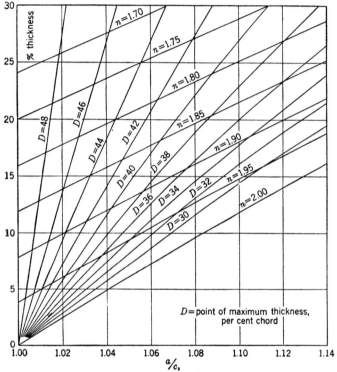

FIG. 6.15.   Parameters for Kármán-Trefftz symmetrical airfoils.

In view of the purely academic interest now given to the Kármán-Trefftz airfoils their mathematical development will not be given here; interested readers may consult Ref. 6.5. The final result of some lengthy but elementary algebra is that the coordinates (ξ, η) of a Kármán-Trefftz airfoil may be found from

$$\xi = nc_1 \frac{\lambda_1{}^2 - 1}{\lambda_1{}^2 - 2\lambda_1 \cos \phi_1 + 1} \tag{6.29}$$

$$\eta = nc_1 \frac{2\lambda_1 \sin \phi_1}{\lambda_1{}^2 - 2\lambda_1 \cos \phi_1 + 1} \tag{6.30}$$

where n is some number slightly less than 2.0 and $k = a/c_1$ is a thickness parameter. (n and k may be determined from Fig. 6.15 for particular airfoils). Other symbols include

$$\phi_1 = n\phi$$

$$\lambda_1 = \lambda^n$$

$$\lambda = \frac{k(1 + bt)\sin\phi}{b + t}$$

$$\phi = \tan^{-1}\frac{b + t}{k(1 + bt) - (1 - bt)}$$

$$b = \tan\frac{\beta}{2}, \text{ a camber parameter}$$

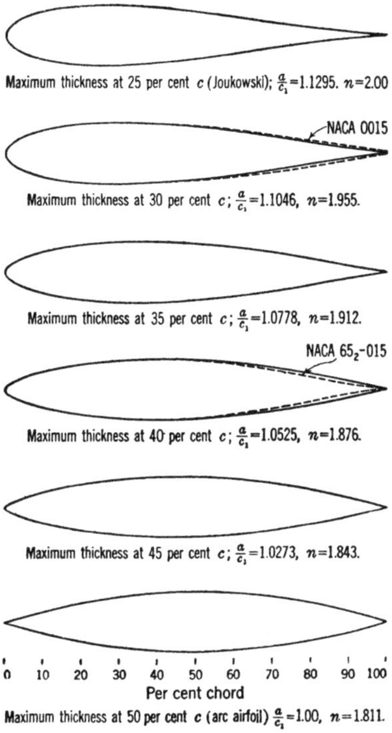

Maximum thickness at 25 per cent c (Joukowski); $\frac{a}{c_1}=1.1295$. $n=2.00$

NACA 0015

Maximum thickness at 30 per cent c; $\frac{a}{c_1}=1.1046$, $n=1.955$.

Maximum thickness at 35 per cent c; $\frac{a}{c_1}=1.0778$, $n=1.912$.

NACA 65_2-015

Maximum thickness at 40 per cent c; $\frac{a}{c_1}=1.0525$, $n=1.876$.

Maximum thickness at 45 per cent c; $\frac{a}{c_1}=1.0273$, $n=1.843$.

| 0 | 10 | 20 | 30 | 40 | 50 | 60 | 70 | 80 | 90 | 100 |

Per cent chord

Maximum thickness at 50 per cent c (arc airfoil) $\frac{a}{c_1}=1.00$, $n=1.811$.

Fig. 6.16. Kármán-Trefftz airfoils 15 per cent thick.

The chord of the Kármán-Trefftz airfoil will be

$$\text{Total chord} = \left[\frac{\left(\dfrac{k}{k-1}\right)^n + 1}{\left(\dfrac{k}{k-1}\right)^n - 1} + 1\right]nc_1 \qquad (6.31)$$

As a matter of general interest a number of Kármán-Trefftz profiles are presented in Fig. 6.16. Comparison between the Kármán-Trefftz

118   *BASIC WING AND AIRFOIL THEORY*

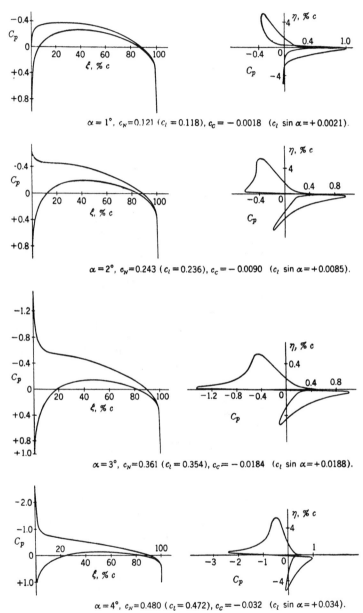

$\alpha = 1°$, $c_N = 0.121$ ($c_l = 0.118$), $c_c = -0.0018$ ($c_l \sin \alpha = +0.0021$).

$\alpha = 2°$, $c_N = 0.243$ ($c_l = 0.236$), $c_c = -0.0090$ ($c_l \sin \alpha = +0.0085$).

$\alpha = 3°$, $c_N = 0.361$ ($c_l = 0.354$), $c_c = -0.0184$ ($c_l \sin \alpha = +0.0188$).

$\alpha = 4°$, $c_N = 0.480$ ($c_l = 0.472$), $c_c = -0.032$ ($c_l \sin \alpha = +0.034$).

Fig. 6.17. Normal and chordwise pressure distribution for Kármán-Trefftz airfoil, 11 per cent thick at 40 per cent chord.   The data in parentheses are from the theoretical formulas and demonstrate the customary order of agreement due to practical integrational difficulties.

airfoil with the maximum thickness at the 30 per cent chord and an NACA 0015 illustrates good agreement over the leading 30 per cent, but the NACA 0015 has a much thicker afterbody. The same applies to the Kármán-Trefftz airfoil with the maximum thickness at the 40 per cent chord and an NACA 65-015. When the maximum thickness is moved to the 50 per cent chord point, the airfoil degenerates into a lenticular (double-convex circular arc) form.

6.7. Pressure Distribution over a Kármán-Trefftz Airfoil. The pressure distribution over a Kármán-Trefftz airfoil follows from the method outlined in the section on Joukowski airfoils. The velocity at the untransformed point P_z is

$$q_z = 2V \sin \theta' + \frac{\Gamma}{2\pi a} \tag{6.32}$$

where θ' and a are defined in Fig. 6.7, and

$$\Gamma = 4\pi a V \sin (\alpha + \beta) \tag{6.33}$$

From Eq. (6.17)

$$q_z = 2V[\sin \theta' + \sin (\alpha + \beta)]$$

and from Eq. (5.9)

$$q_\zeta = \frac{|q_z|}{\left|\dfrac{d\zeta}{dz}\right|}$$

The relation of Eq. (6.28) may be written (Ref. 6.5)

$$\frac{d\zeta}{dz} = n^2 \frac{\lambda_1}{\lambda} \frac{1 + \lambda^2 - 2\lambda \cos \phi}{1 + \lambda_1^2 - 2\lambda_1 \cos \phi_1} \tag{6.34}$$

and the pressure distribution then found from Eq. (3.10).

As a matter of general interest, several pressure plots are shown in Fig. 6.17 for an 11 per cent thick Kármán-Trefftz airfoil with the maximum thickness at the 40 per cent chord point. It will be noted that the sharper leading edge yields more favorable pressure gradients at very low angles of attack than those of the simple Joukowski. In the practical case, however, the stall may come a little earlier, with resulting lower maximum lift.

REFERENCES

6.1. E. Trefftz, Graphic Construction of Joukowski Wings, *TM* 336, 1925.
6.2. H. Glauert, A Generalized Type of Joukowski Aerofoil, *R & M* 911, 1924.
6.3. Muriel Glauert, Two-dimensional Aerofoil Theory, *JRAS*, 1923.
6.4. N. A. V. Piercy, "Aerodynamics," p. 178, English University Press, London.
6.5. Alan Pope, Wing and Airfoil Lecture Notes, Georgia Institute of Technology, Atlanta, Ga.

CHAPTER 7

THIN-AIRFOIL THEORY

Until the present time we have been concerned with transformations that yield the field of flow about an airfoil so that the characteristics of the airfoil (the angle of zero lift, the lift and pitching moment, the pressure distribution, and the flow pattern about the airfoil) may be determined. In practice, two other processes come under scrutiny: (1) finding the characteristics of an airfoil already drawn or in use; (2) finding an airfoil that will develop certain characteristics. The solutions to both these problems come directly from the theory so far discussed. In this chapter and the one following we shall consider methods for determining the characteristics of airfoils whose dimensions are already known.

It was shown in Chap. 5 that thickness contributes very little to the lift and moment of an airfoil. This fact plus the convenience of unextended mathematical steps led early aerodynamicists to consider thick airfoils as consisting of their mean lines only for the purpose of analysis. The theory which resulted is called the *Munk-Glauert-Birnbaum thin-airfoil theory*, after those who made contributions to it. It yields in a short time the angle of zero lift (which indirectly determines the lift at any angle since the lift curve slope for a thin airfoil is 2π per radian), the moment coefficient about the quarter chord, and the ideal angle. (It is general practice in thin-airfoil theory to assume that the aerodynamic center of an airfoil is at the quarter chord, and that $c_{mac} = c_{m\frac{1}{4}}$.)

Extensions to this theory by Jacobs, Theodorsen, Allen, Abbott, von Doenhoff, and Stivers make possible a number of different methods by which the characteristics of thin *and* thick airfoils may be determined.

7.1. The Thin Airfoil. Let a thick airfoil be replaced by its mean line, and let the mean line be described by

$$\zeta = \xi + i\eta \tag{7.1}$$

Place it on the ξ axis in the ζ plane so that it extends from $+2a$ to $-2a$. Since a flat plate when so disposed yields a circle when transformed by the Joukowski relation, it is reasonable that an airfoil mean line which differs little from a flat plate would when similarly treated yield a near, or pseudo, circle. Let us call the circle

$$z = ae^{i\phi} \tag{7.2}$$

and the near circle

$$z' = a(1 + r)e^{i\theta} \qquad (7.3)$$

It is seen that r varies and hence corresponds to the difference between the radius of the circle and of the near circle. The point (ξ, η) on the airfoil transforms by the Joukowski relation to the point z' on the near circle. It then transforms (by a much more complex transformation that

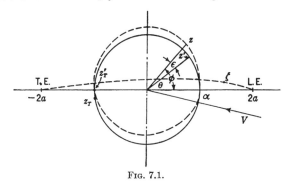

Fig. 7.1.

we must yet determine) to the point z on the circle. We may link the two arguments by

$$\epsilon = \phi - \theta$$

and both ϵ and r are small.

The relation between the near circle and the airfoil is

$$\zeta = z' + \frac{a^2}{z'} \qquad (7.4)$$

and that between the near circle and the circle is (see Appendix 1)

$$z' = z\left(1 + i \sum_1^\infty A_n \frac{a^n}{z^n}\right) \qquad (7.5)$$

so that (again see Appendix 1)

$$\epsilon = -\sum_1^\infty A_n \cos n\theta \qquad (7.6)$$

and

$$r = \sum_1^\infty A_n \sin n\theta \qquad (7.7)$$

From page 299 of Ref. 7.2 the constants may be found from

$$A_n = \frac{2}{\pi} \int_0^\pi r \sin n\theta \, d\theta \qquad (7.8)$$

7.2. Lift of a Thin Airfoil. Let us first employ the Kutta-Joukowski equation to find the lift of a thin airfoil. From Eq. (4.51) the potential function of a circular cylinder with circulation in a flow inclined at angle α is

$$w = -Vze^{i\alpha} - \frac{Va^2e^{-i\alpha}}{z} - \frac{i\Gamma}{2\pi}\ln\frac{z}{a}$$

and according to our hypothesis (Sect. 5.4) the rear stagnation point of the circle where $q_z = 0$ will be at the point $\phi = \theta + \epsilon_T = \pi + \epsilon_T$.

At the surface of the circular cylinder

$$\frac{dw}{dz} = -Ve^{i\alpha} + \frac{Va^2e^{-i\alpha}}{z^2} - i\frac{\Gamma}{2\pi z}$$

$$= -2iVe^{-i\phi}\left[\sin(\alpha + \phi) + \frac{\Gamma}{4\pi aV}\right] \tag{7.9}$$

and this becomes zero at $\phi = \pi + \epsilon_T$ if

$$\Gamma = 4\pi aV \sin(\alpha + \epsilon_T) \tag{7.10}$$

Thus the value of ϵ at the trailing edge $\epsilon_T = -\alpha_{ZL}$ and the lift is, from Eq. (4.24),

$$L = \frac{\rho}{2}V^2 \cdot 4a \cdot 2\pi \sin(\alpha + \epsilon_T) \tag{7.11}$$

7.3. Velocity over a Thin Airfoil. From Eqs. (7.9) and (7.10) we write the velocity at point ϕ in the z plane as

$$q_z = \left|\frac{dw}{dz}\right| = 2V[\sin(\alpha + \phi) + \sin(\alpha + \epsilon_T)] \tag{7.12}$$

To find q_ζ we must, according to Eq. (5.9), know $\dfrac{d\zeta}{dz}$. The procedure follows the analysis for Eq. (5.21).

Substituting from $\phi = \theta + \epsilon$, we get

$$q_z = 2V(\alpha\cos\theta\cos\epsilon - \alpha\sin\theta\sin\epsilon + \sin\theta\cos\epsilon + \sin\epsilon\cos\theta + \alpha + \epsilon_T)$$

which becomes ($\cos\epsilon \cong 1.0$, $\sin\theta\sin\epsilon \cong 0$)

$$q_z = 2V(\sin\theta + \alpha\cos\theta + \epsilon\cos\theta + \alpha + \epsilon_T) \tag{7.13}$$

Differentiating Eq. (7.5), we have

$$\frac{dz'}{dz} = 1 - i\sum_{1}^{\infty}A_n(n-1)\frac{a^n}{z^n}$$

which upon substitution from $z = ae^{i\phi}$ and $\phi = \theta + \epsilon$ becomes

$$\frac{dz'}{dz} = 1 - i \sum_{1}^{\infty} A_n(n - 1)e^{-in\theta}$$

if we consider $e^{in\epsilon} \cong 1.0$.

The absolute value of $\dfrac{dz'}{dz}$ is

$$\left|\frac{dz'}{dz}\right| = \sqrt{\left(1 - \sum_{1}^{\infty} A_n(n - 1) \sin n\theta\right)^2 + \left(\sum_{1}^{\infty} A_n(n - 1) \cos n\theta\right)^2}$$

Neglecting the square terms that ensue since they are essentially ϵ^2 and r^2 and expanding the remainder by the binomial theorem, we have, finally,

$$\left|\frac{dz'}{dz}\right| = 1 - \sum_{1}^{\infty} nA_n \sin n\theta + \sum_{1}^{\infty} A_n \sin n\theta$$

$$= 1 - \frac{d\epsilon}{d\theta} + r \tag{7.14}$$

We may find $\dfrac{d\zeta}{dz'}$ by differentiating Eq. (7.4). Thus

$$\frac{d\zeta}{dz'} = 1 - \frac{a^2}{(z')^2}$$

Substituting $z' = a(1 + r)e^{i\theta}$ and neglecting r^2 as small, we get

$$\frac{d\zeta}{dz'} = 1 - (1 - 2r)e^{-2i\theta}$$

Expanding $e^{-2i\theta}$ into sines and cosines, we get

$$\frac{d\zeta}{dz'} = 2e^{-i\theta}(i \sin \theta + re^{-i\theta})$$

whose absolute value is

$$\left|\frac{d\zeta}{dz'}\right| = 2\sqrt{r^2 \cos^2 \theta + (1 - r)^2 \sin^2 \theta}$$

$$= 2 \sin \theta (1 - 2r)^{\frac{1}{2}}$$

Upon expanding and neglecting terms of r^2 or higher, this becomes

$$\left|\frac{d\zeta}{dz'}\right| = 2 \sin \theta (1 - r) \tag{7.15}$$

Combining Eqs. (7.14) and (7.15), we get

$$\left|\frac{d\zeta}{dz}\right| = \left|\frac{d\zeta}{dz'}\right|\left|\frac{dz'}{dz}\right| = 2\sin\theta\left(1 - r^2 - \frac{d\epsilon}{d\theta}\right) + r\frac{d\epsilon}{d\theta}$$

From actual examples $\frac{d\epsilon}{d\theta}$ is moderate in size so that when multiplied by r the product is negligible. Finally

$$\left|\frac{d\zeta}{dz}\right| = 2\sin\theta\left(1 - \frac{d\epsilon}{d\theta}\right)^* \tag{7.16}$$

Hence, from Eqs. (7.13) and (7.16) the velocity at any point θ on the airfoil is

$$q_\zeta = V\left[1 + (\alpha + \epsilon)\cot\theta + (\alpha + \epsilon_T)\csc\theta + \frac{d\epsilon}{d\theta}\right] \tag{7.17}$$

7.4. Pressure over a Thin Airfoil. The increase of velocity at point P on the upper surface of a thin airfoil will be the same as the decrease in velocity on the lower surface at point P. Writing Bernoulli's equation for upper and lower surfaces relative to the free-stream static pressure, we have

$$p_u - p_0 = \frac{\rho}{2}V^2 - \frac{\rho}{2}(V + \Delta V)^2$$

and

$$p_l - p_0 = \frac{\rho}{2}V^2 - \frac{\rho}{2}(V - \Delta V)^2$$

The total pressure coefficient

$$P = \frac{p_l - p_u}{\frac{\rho}{2}V^2} = 4\frac{\Delta V}{V} \tag{7.18}$$

From Eqs. (7.18) and (7.17) (since $\Delta V = q_\zeta - V$)

$$P = 4\left(\alpha\frac{1 + \cos\theta}{\sin\theta} + \frac{\epsilon_T + \epsilon\cos\theta}{\sin\theta} + \frac{d\epsilon}{d\theta}\right) \tag{7.19}$$

7.5. Lift and Moment Coefficients. The lift coefficient of a thin airfoil can be found directly from Eq. (7.11) as

$$c_l = 2\pi(\alpha + \epsilon_T) \tag{7.20}$$

At this point we shall make an important change of symbols. Since airfoil coordinates are most frequently given in fractions of the chord

---

* Hence $d\zeta/dz = 0$ at $\theta = \pi$, and a finite velocity is possible at the trailing edge of the airfoil. A study of Eq. (7.17) reveals that the trailing-edge velocity is $V_{\text{free stream}}$.

using x and y for abscissa and ordinate, we shall now write

$$x = \frac{\xi}{c} \tag{7.21}$$

$$y = \frac{\eta}{c} \tag{7.22}$$

If in addition we make

$$\cos \theta = 1 - 2x \tag{7.23}$$

so that

$$\sin \theta = 2 \sqrt{x(1 - x)} \tag{7.24}$$

the airfoil will be turned around so that $x = 0$ will be its leading edge instead of the half-chord point. This change (see Fig. 7.2) does not in

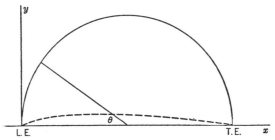

FIG. 7.2. The thin airfoil on the xy axes.

any way invalidate the relations we have already derived in which θ appears since particular chord stations still correspond to the same θ.

The moment about the leading edge is then

$$M_{le} = -\frac{\rho}{2} V^2 \int_0^{1.0} Px \, dx$$

and upon substituting from Eqs. (7.19) and (7.23) and integrating we have

$$c_{mle} = -\frac{\pi}{2} \alpha - \pi \epsilon_T + \frac{\pi}{4} A_2$$

Solving Eq. (7.20) for α, substituting into the equation above, and then transferring to the quarter chord, we have

$$c_{m\frac{1}{4}} = \frac{\pi}{4} A_2 - \frac{\pi \epsilon_T}{2} \tag{7.25}$$

The presentation of Eqs. (7.20) and (7.25) is not of much use, however, until we add methods of evaluating A_2 and ϵ_T. Glauert in Ref. 7.1 introduced

$$\mu_0 = \frac{\pi}{8} A_2 \tag{7.26}$$

so that

$$c_{m\frac{1}{4}} = 2\mu_0 - \frac{\pi}{2}\epsilon_T \tag{7.27}$$

Equation (7.3) may be expanded using Eq. (7.1) to get

$$\zeta = 2a(\cos\theta + ir\sin\theta) \tag{7.28}$$

from which

$$\xi = 2a\cos\theta \tag{7.29}$$

and

$$\eta = 2ar\sin\theta \tag{7.30}$$

The value of A_2 comes directly from Eq. (7.8) as

$$A_2 = \frac{2}{\pi}\int_0^\pi r\sin 2\theta\, d\theta$$

and since $r = \eta/(2a\sin\theta)$ from Eq. (7.30) we have

$$A_2 = \frac{2}{\pi}\int_0^\pi \frac{\eta}{a}\cos\theta\, d\theta \tag{7.31}$$

or

$$\mu_0 = \frac{1}{4a}\int_0^\pi \eta\cos\theta\, d\theta \tag{7.32}$$

The value of ϵ_T is the value of ϵ when $\theta = \pi$, or from Eq. (7.6)

$$\epsilon_T = A_1 - A_2 + A_3 - \cdots$$

which by Eq. (7.8) is

$$\epsilon_T = \frac{2}{\pi}\int_0^\pi (A_1\sin\theta + A_2\sin 2\theta + \cdots)(\sin\theta - \sin 2\theta + \cdots)\, d\theta \tag{7.33}$$

Equation (7.33) is the same as

$$\epsilon_T = I\frac{2}{\pi}\int_0^\pi r(e^{i\theta} - e^{2i\theta} + \cdots)\, d\theta$$

where $I =$ imaginary part.
Hence

$$\epsilon_T = I\frac{2}{\pi}\int_0^\pi \frac{re^{i\theta}}{1 + e^{i\theta}}\, d\theta = \frac{1}{\pi}\int_0^\pi \frac{r\sin\theta}{1 + \cos\theta}\, d\theta \tag{7.34}$$

and thus, substituting from Eq. (7.30),

$$\epsilon_T = \frac{1}{2\pi a}\int_0^\pi \frac{\eta}{1 + \cos\theta}\, d\theta \tag{7.35}$$

In the nondimensional plane μ_0 and ϵ_T become

$$\epsilon_T = \int_0^1 y f_1(x) \, dx \tag{7.36}$$

where

$$f_1(x) = \frac{1}{\pi(1-x)(\sqrt{x(1-x)})} \tag{7.37}$$

and

$$\mu_0 = \int_0^1 y f_2(x) \, dx \tag{7.38}$$

when

$$f_2(x) = \frac{1-2x}{\sqrt{x(1-x)}} \tag{7.39}$$

The convenience of this method becomes evident when it is noted that $f_1(x)$ and $f_2(x)$ are functions of selected chord positions only and need be calculated only once (see Table 7.1). However, special care must be taken at $x = 0$ and 1.0, where $f_1(x)$, $f_2(x)$ and, $f_3(x)$ all become infinite.

TABLE 7.1. VALUES OF $f_1(x)$, $f_2(x)$, AND $f_3(x)$

x	$f_1(x)$	$f_2(x)$	$f_3(x)$
0	∞	∞	∞
0.025	2.09	6.08	39.73
0.05	1.54	4.13	13.78
0.10	1.18	2.67	4.72
0.15	1.05	1.96	2.45
0.20	0.99	1.50	1.49
0.25	0.98	1.15	0.972
0.30	0.99	0.87	0.662
0.35	1.03	0.63	0.442
0.40	1.08	0.41	0.271
0.45	1.16	0.20	0.130
0.50	1.27	0.00	0
0.55	1.42	−0.20	−0.130
0.60	1.62	−0.41	−0.271
0.65	1.91	−0.63	−0.442
0.70	2.31	−0.87	−0.662
0.75	2.94	−1.15	−0.972
0.80	3.98	−1.50	−1.49
0.85	5.94	−1.96	−2.45
0.90	10.62	−2.67	−4.72
0.95	29.24	−4.13	−13.78
1.00	∞	−∞	−∞

Considering $f_1(x)$ first at $x = 0$, we differentiate numerator and denominator of $yf_1(x)$ and solve for the limiting value as follows:

$$yf_1(x)_{x=0} = \lim_{x \to 0} \left\{ \frac{1}{\pi} \frac{2[x(1 - x)]^{\frac{1}{2}}}{1 - 5x + 4x^2} \frac{dy}{dx} \right\}$$

and, putting in the limit $x = 0$, we get

$$yf_1(x)_{x=0} = 0$$

except for $dy/dx = \infty$, which may be disregarded.

The value at $x = 1$ requires a different approach. It is reasonable to assume that the rear 5 per cent of the camber line is a straight line. Its equation is, then,

$$y = 20y_1(1 - x) \tag{7.40}$$

where y_1 = value of y at $x = 0.95$.

Using Eq. (7.40), we find the contribution of the last 5 per cent of the camber line to ϵ_T, as follows:

$$\Delta\epsilon_T = \frac{1}{\pi} \int_{0.95}^{1.00} \frac{20y_1 (1 - x)}{(1 - x) \sqrt{x(1 - x)}} \, dx \tag{7.41}$$

$$= \frac{20y_1}{\pi} \left[\cos^{-1} (1 - 2x) \right]_{0.95}^{1.00}$$

$$= 2.87y_1$$

If we had assumed the rearward camber line to be parabolic, $\Delta\epsilon_T = 2.92y_1$ and it hence appears reasonable to use the mean value of $2.9y_1$ for calculations.

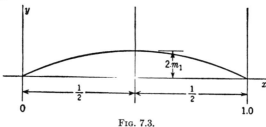

FIG. 7.3.

The treatment of $f_2(x)$ is left to the student.

Now then, if the equation of the camber line is known, Eqs. (7.27), (7.36), and (7.38) may be used to find α_{ZL} and $c_{m\frac{1}{4}}$. Although this is rarely the case, three particular examples come to mind:

1. *The Flat-plate Airfoil.* In this case $y = 0$, and Eqs. (7.36), (7.38), and (7.27) yield $\alpha_{ZL} = 0$, $\mu_0 = 0$, and $c_{m\frac{1}{4}} = 0$ directly.

2. *The Parabolic-camber Airfoil.* Placing a parabolic-camber thin airfoil on the axes as in Fig. 7.3, we have

$$y = 8m_1x(1 - x) \qquad (7.42)$$

and

$$\alpha_{ZL} = -4m_1$$
$$\mu_0 = 0$$

and

$$c_{m\frac{1}{4}} = -2\pi m_1$$

(Note that $m_1 = m/t$ as used in Chap. 5.)

It is seen that the moment is the same as that of a circular-arc airfoil [Eq. (5.52)] within the limits of our assumption.

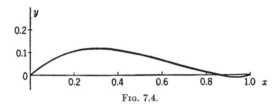

FIG. 7.4.

3. *The Constant-center-of-pressure Airfoil.* Consider a reflex-camber airfoil as in Fig. 7.4, described by

$$y = px(1 - x)(q - x)$$

where p = maximum camber and moment parameter
q = chordwise location of start of reflex

We have then

$$\alpha_{ZL} = -\frac{p}{8}(4q - 3)$$

$$\mu_0 = \frac{\pi p}{64}$$

$$c_{m\frac{1}{4}} = \frac{\pi p}{32}(7 - 8q)$$

We see that a small reflex exerts a powerful effect on the chord loading, and indeed, when $q = \frac{7}{8}$, the moment coefficient vanishes. It has already been demonstrated that no moment about the quarter chord is contributed by the additional lift due to angle of attack, and thus an airfoil with $q = \frac{7}{8}$ must have a constant center of pressure located at the quarter chord as does a symmetrical airfoil.

4. *The Arbitrary Thin Airfoil.* For the general case the camber line is usually not simply expressed as a function of x, and we must resort to a

graphical integration.   That is, to find α_{ZL} and $c_{m\frac{1}{4}}$ for an arbitrary airfoil, we must find $yf_1(x)$ and $yf_2(x)$ and get the areas under the plots of
these two functions against x.   The $yf_1(x)$ curve is integrated from 0 to
0.95 and upon the addition of $2.9y_1$ yields α_{ZL}.   The μ_0 curve is integrated from 0 to 1.0 and with the above value of ϵ_T yields $c_{m\frac{1}{4}}$ by Eq.
(7.27).

Example 7.1.   By means of the thin-airfoil theory find the angle of zero lift and
$c_{m\frac{1}{4}}$ for the modified NACA 65,311 airfoil shown in Fig. 7.5.   Ordinates for this

Fig. 7.5.   Modified NACA 65,311 airfoil.   (*Courtesy of Journal of the Aeronautical Sciences.*)

airfoil are given in Table 7.2.   (The modification embraced thickening the airfoil
over the last 30 per cent of the chord to avoid the unsatisfactory control-surface
hinge moments and construction difficulties usually associated with a cusped
trailing edge.   In this case no change of minimum drag was realized, but the lift
curve slope was slightly decreased.)

TABLE 7.2.   ORDINATES OF 65,311 (MODIFIED)

Station	Upper surface per cent chord	Lower surface per cent chord
0.0	0.0	0
1.25	1.295	1.223
2.50	1.187	1.511
5.00	2.807	1.942
7.50	3.596	2.230
10.00	4.245	2.519
15.00	5.25	2.970
20.00	5.97	3.239
30.00	6.98	3.527
40.00	7.48	3.527
50.00	7.41	3.383
60.00	6.76	3.022
70.00	5.61	2.519
80.00	4.172	1.870
90.00	2.375	1.079
95.00	1.295	0.648
97.50	0.79	0.40
100.00	0.16	0.16

TABLE 7.3

x	y_u	y_l	$\frac{1}{2}(y_u - y_l)$	$yf_1(x)$	$yf_2(x)$
0	0	0	0	0	0
0.0125	0.01295	0.01223	0.000360	0.0001045	0.003165
0.0250	0.01871	0.01511	0.00180	0.00377	0.01095
0.0500	0.02807	0.01942	0.004325	0.00666	0.01785
0.0750	0.03506	0.2230	0.00683	0.00894	0.02210
0.1000	0.04245	0.2519	0.00863	0.01019	0.02305
0.1500	0.0525	0.0297	0.01140	0.01198	0.02235
0.2000	0.0597	0.03239	0.01365	0.01350	0.02050
0.3000	0.0698	0.03527	0.01726	0.01710	0.0150
0.4000	0.0748	0.03527	0.01977	0.02135	0.0081
0.5000	0.0741	0.03383	0.02014	0.02560	0.0000
0.6000	0.0676	0.03022	0.01869	0.03025	-0.00766
0.7000	0.0561	0.02519	0.0154	0.0356	-0.01338
0.8000	0.04172	0.0187	0.0115	0.0457	-0.01723
0.9000	0.02375	0.01079	0.00648	0.0688	-0.01730
0.9500	0.01295	0.00648	0.00323	0.0945	-0.01332
0.9750	0.0079	0.0040	0.0019	—	-0.01152
1.0000	0.0016	0.0016	0.000	—	0

The calculations for this airfoil are shown in Table 7.3, and the plots of $yf_1(x)$ and $yf_2(x)$ needed to get ϵ_T and μ_0 are shown in Figs. 7.6a and b. From them we get

$$-\epsilon_T = \alpha_{ZL} = -[0.02730 + 2.9(0.00323)](57.3)$$
$$= -2.13°$$

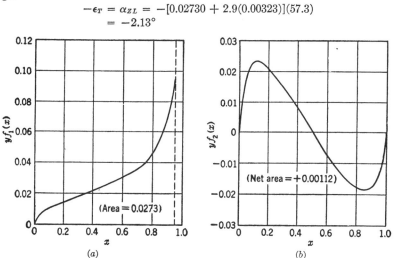

(a)   (b)

FIG. 7.6.   (a) Plot of $yf_1(x)$ vs. x for determination of α_{ZL} (example airfoil).   (b) Plot of $yf_2(x)$ vs. x for determination of μ_0 (example airfoil).

132   BASIC WING AND AIRFOIL THEORY

and

$$c_{m\frac{1}{4}} = 2(0.00112) - \frac{\pi}{2}(0.03667)$$
$$= -0.0554$$

These values compare with $-2.5°$ and -0.060 obtained by wind-tunnel tests.

A correction not necessary in this case must be noted in order that the process be general. The thin-airfoil theory uses the camber referred to the line joining the leading and trailing edges of the airfoil. For most modern airfoils this is also the chord, but in case some other chord line is in use the value of α_{ZL} from thin-airfoil theory must be corrected by the angular difference between chord lines in order to have a common reference.

Results from similar calculations for several other airfoils are presented in the tabulation.

Airfoil	α_{ZL} Theory	α_{ZL} Practice	$c_{m\frac{1}{4}}$ Theory	$c_{m\frac{1}{4}}$ Practice
65,209	−1.51	−1.30	−0.040	−0.035
65,410	−2.62	−2.40	−0.066	−0.065
64,A-212	−1.59	−1.80	−0.043	−0.040
747A315	−1.11	−1.10	−0.009	−0.012
63(420)-517	−3.94	−3.20	−0.099	−0.090

7.6. The Ideal Angle of Attack. Now we are familiar with the use of the Kutta condition and, indeed, have employed it to find the proper circulation for our thin airfoil in order to have the flow leave the trailing edge smoothly. Equally important for efficient operation of our thin airfoil is that the flow enter the leading edge smoothly. This condition (first discussed by Theodorsen in Ref. 7.5) requires that the pressure coefficient at the leading edge be zero.

Inserting $\cos \theta = 1 - 2x$, $\sin^2 \theta = 4x(1 - x)$ into Eq. (7.19), we get

$$P = 4\frac{d\epsilon}{dx}\sqrt{x(1 - x)} + \frac{2}{\sqrt{x(1 - x)}}[2\alpha(1 - x) + \epsilon_T + \epsilon(1 - 2x)]$$

which can be written

$$P = 4(\alpha + \epsilon_T)\sqrt{\frac{1 - x}{x}} + 4\frac{d}{dx}[(\epsilon - \epsilon_T)\sqrt{x(1 - x)}] \quad (7.43)$$

Writing Eq. (7.43) for a point only Δx from the leading edge, we have, letting $\epsilon = \epsilon_N$ (N = nose), $1 - \Delta x \cong 1.0$, and Δx^2 be negligible,

$$\frac{P_{\Delta x}}{4} = (\alpha + \epsilon_T)\sqrt{\frac{1}{\Delta x}} - \frac{d}{dx}[(\epsilon_N - \epsilon_T)\sqrt{\Delta x}]$$

Performing the differentiation and neglecting $\Delta x \cdot \dfrac{d\epsilon}{d\alpha}$ as small, we have

$$\frac{P_{\Delta x}}{4} = \frac{1}{\sqrt{\Delta x}}\left(\alpha + \frac{\epsilon_T + \epsilon_N}{2}\right)$$

which is zero only when the ideal angle of attack

$$\alpha_I = -\frac{\epsilon_T + \epsilon_N}{2} \tag{7.44}$$

Theodorsen, using an approach analogous to our Sect. 7.5, found the ideal angle to be (page 9 of Ref. 7.5)

$$\alpha_I = \int_0^{1.0} y f_3(x)\ dx*$$

where

$$f_3(x) = \frac{1 - 2x}{2\pi[x(1 - x)]^{\frac{3}{2}}} \quad \text{(see Table 7.1)}$$

A shorter form is the approximation that

$$\alpha_I = 623(y_1 - y_5) + 47(y_2 - y_4) \tag{7.45}$$

where y_1, y_2, y_4, and y_5 are the ordinates of the mean camber line at $x = 0.542, 12.574, 87.426,$ and 99.458 per cent chord and α_I is the ideal angle in degrees.

At all angles other than the ideal we find infinite velocities near the leading edge as the flow tries to negotiate the sharp nose, and the thin-airfoil theory is acceptable only because actual airfoils have finite leading-edge radii which provide reasonable velocities. We shall continue our discussion of the ideal angle later.

It will be noted that we have been concerned only with the values of ϵ at the leading and trailing edges of the airfoil. The entire ϵ curve may be plotted, if desired, by a method given by Theodorsen in Ref. 7.5. In general, ϵ is negative over the first third of an airfoil, and it usually varies numerically in a smooth curve between -0.1 and 0.1.

7.7. Thin Airfoils with Trailing-edge Flaps. The theory of thin airfoils has been applied (Refs. 7.9, 7.10, 7.11) to the case of airfoils having trailing-edge flaps with a good degree of success. The basic approach, since we are most interested in the contribution of the flap, is to consider the forward part of the airfoil as uncambered. The complete airfoil then

* Abbott and von Doenhoff in Ref. 8.10 discuss the special treatment required for the above integral in order to avoid difficulties at the leading and trailing edges.

consists of a section AB (see Fig. 7.7) from leading edge to hinge
point, plus the flap of chord Ec, where E is some number usually less
than 0.5 and c the chord of the airfoil including flap. The previously
outlined integrations are carried out in two parts, first from the leading
edge to the hinge point, and then from there to the trailing edge. The
moment of the flap about the hinge point is called the *flap hinge moment*

FIG. 7.7.

and is of interest both from a load
standpoint and to furnish data for
the flap power-drive system, if any.
Although the effects of flaps are of
interest for both controls and lift,
they will be omitted here in the interest of brevity. A problem concerning trailing-edge flaps is included at the end of this chapter.

7.8. Total Pressure Distribution over a Thin Airfoil. The characteristics of a thin airfoil may be developed from an entirely different approach
by assuming that the airfoil may be represented by a series of vortices
distributed along the mean line whose strength is that needed to induce
the same velocities that the airfoil develops. Any individual vortex will
have the strength γ, square foot per second per foot of chord, and the
total circulation will be

$$\Gamma = \int_0^c \gamma \, dx$$

where c = the chord.
Values of γ may be found through the use of the condition that the flow
must be everywhere tangent to the mean line, or, mathematically,

$$\frac{dy}{dx} = \alpha + \frac{v_n}{V}$$

where v_n = local vertical induced velocity.
This method is discussed by Glauert in Ref. 7.8 and yields in the final
analysis the same results as our transformation approach.

So far we have considered only the process needed to get the ideal angle,
the angle of zero lift, and the moment of an airfoil. Obviously, an important additional item desired is the pressure distribution over the entire
airfoil. Equation (7.19) could be used to get it, but this would entail a
long and involved procedure indeed. Besides, from a practical standpoint, it would be desirable to have a method that includes the necessary
empirical corrections so that it could be used directly for aircraft structural loadings. Such a process has been developed by Jacobs (Ref. 7.3)
based upon the earlier work of Theodorsen. Jacobs found it more convenient to introduce the empirical corrections which are necessary if the
pressure distribution were divided into that at zero lift plus an additional

part instead of that at the ideal angle plus an additional part. This in no way disturbs our logic, as it is quite easy to lump a little amount of additional lift in with the ideal and consider them together.

We shall start by first getting the pressure distribution over a flat plate. Consider Eq. (4.17),

$$q_z = 2V \sin \theta' + \frac{\Gamma}{2\pi a}$$

For a flat plate Eq. (5.17) gives

$$\Gamma = 4\pi a V \alpha$$

and since

$$\frac{d\zeta}{dz} = 1 - \frac{c_1{}^2}{z^2} = 2 \sin \theta \tag{7.46}$$

we get

$$q_\zeta = \frac{2V\alpha \cos \theta + 2V \sin \theta + 2V\alpha}{2 \sin \theta} \tag{7.47}$$

and

$$\frac{\Delta V}{V} = \frac{q_\zeta - V}{V} = \frac{\alpha(1 + \cos \theta)}{\sin \theta}$$

It was through recognition of this quantity in Eq. (7.19) that the concept of adding the flat-plate pressure distribution to another distribution (that at the ideal angle) first developed.

Through Eq. (7.18) we now have

$$P = \frac{4\alpha(1 + \cos \theta)}{\sin \theta} = 4\alpha \cot \frac{\theta}{2} \tag{7.48}$$

for a flat plate.

The part of the pressure distribution due to the camber we shall write as

$$P = 4 \sum_1^\infty B_n \sin n\theta \tag{7.49}$$

getting for the total

$$P = 4 \left[\frac{\alpha(1 + \cos \theta)}{\sin \theta} + \sum_1^\infty B_n \sin n\theta \right] \tag{7.50}$$

and since

$$c_l = 2\pi(\alpha + \epsilon_T)$$

$$P = \frac{2c_l}{\pi} \frac{1 + \cos \theta}{\sin \theta} + \left[-4\epsilon_T \frac{(1 + \cos \theta)}{\sin \theta} + 4 \sum_1^\infty B_n \sin n\theta \right] \tag{7.51}$$

When $c_l = 0$, the so-called *zero-lift pressure distribution* is obtained, represented by

$$P_{ZL} = P_0 = -4\epsilon_T \frac{(1 + \cos \theta)}{\sin \theta} + 4 \sum_1^\infty B_n \sin n\theta \qquad (7.52)$$

If we call the first term of Eq. (7.51) the *additional pressure distribution due to lift* $c_l P_a$, we have

$$P_a = \frac{2}{\pi} \frac{1 + \cos \theta}{\sin \theta} \qquad (7.53)$$

Using $x = (1 - \cos \theta)/2$, we may select values of x and find the theoretical values for P_a. These are shown in Table 7.4 along with practical variations found necessary and explained later.

TABLE 7.4.   ADDITIONAL CHORD LOAD DISTRIBUTION FACTORS FOR AIRFOIL SECTIONS

Sta. per cent chord	P_{ac}	P_{a1}				
		Class B	Class C	Class D	Class E	Theory
0	0	0	0	0	0	∞
1.25	3.2	5.93	4.98	4.32	3.87	5.661
2.5	4.5	4.37	4.23	4.02	3.68	3.977
5.0	5.5	3.20	3.22	3.25	3.27	2.774
7.5	5.9	2.63	2.68	2.76	2.81	2.236
10.0	5.7	2.26	2.32	2.39	2.44	1.910
15.0	5.0	1.77	1.85	1.90	1.95	1.516
20.0	4.3	1.47	1.54	1.58	1.62	1.273
30.0	2.9	1.10	1.14	1.16	1.18	0.9727
40.0	1.4	0.86	0.87	0.88	0.89	0.7798
50.0	0.0	0.67	0.68	0.68	0.69	0.6366
60.0	−1.4	0.51	0.51	0.51	0.51	0.5198
70.0	−2.9	0.38	0.37	0.37	0.36	0.4168
80.0	−4.3	0.25	0.24	0.24	0.23	0.3183
90.0	−5.7	0.13	0.12	0.12	0.11	0.2122
95.0	−5.5	0.06	0.06	0.06	0.06	0.1460
100.0	0.0	0	0	0	0	0

Note. Class A distributions have not been determined.

The total pressure distribution is

$$P = P_0 + c_l P_a \qquad (7.54)$$

Now, referring to Eq. (7.52), we see that the first term of the zero-lift pressure distribution P_0 is proportional to the additional pressure distribution. Hence

$$P_0 = -2\pi\epsilon_T P_a + 4\sum_1^\infty B_n \sin n\theta \tag{7.55}$$

The part of the zero-lift pressure distribution not proportional to the additional pressure distribution is called the *basic pressure distribution* P_b. It is, as will be later demonstrated, the pressure distribution that occurs at the ideal angle of attack. Hence

$$P_b = 4\sum_1^\infty B_n \sin n\theta \tag{7.56}$$

The constants B_n may be evaluated as follows:

The lift of the thin airfoil may be written as (let the chord $= 1.0$)

$$L = \int_0^{1.0} P \cdot \frac{\rho}{2} V^2 \cdot dx \tag{7.57}$$

$$c_l = \int_0^\pi 2\alpha(1 + \cos\theta)\, d\theta + \int_0^\pi \sum_1^\infty \frac{B_n \sin n\theta \sin \theta\, d\theta}{2}$$

And since the second integral exists only for $n = 1$, we have

$$c_l = 2\pi\left(\alpha + \frac{B_1}{2}\right) \tag{7.58}$$

so that

$$\frac{B_1}{2} = \epsilon_T = -\alpha_{ZL} \tag{7.59}$$

The moment about the leading edge is

$$M_{le} = -\frac{\rho}{2} V^2 \int_0^{1.0} Px\, dx$$

$$= -\frac{\rho}{2} V^2 \int_0^{1.0} \left[\alpha(1 - \cos^2\theta) + \sum_1^\infty B_n \sin n\theta \sin \theta\right.$$

$$\left. - \sum_1^\infty B_n \sin n\theta \cos \theta \sin \theta\right] d\theta$$

Now when $n = 1$, the second term in the bracket becomes $\sin^2\theta\, d\theta$ and integrates as before to $\pi/2$, while the last term becomes $\sin^2\theta \cos\theta\, d\theta$ and goes to zero, from 0 to π.

When $n = 2$, the second term becomes zero and the third term is integrable.   We replace the $\cos^2 \theta$ in the parentheses of the first term by $\frac{1}{2} + \frac{\cos 2\theta}{2}$, and integrating

$$M_{le} = -\frac{\rho}{2} V^2 \left(\frac{\pi\alpha}{2} + \frac{\pi}{2} B_1 - \frac{\pi}{4} B_2 \right)$$

from which (again)

$$c_{m_{\frac{1}{4}}} = \frac{\pi}{4} (B_2 - B_1) \qquad (7.60)$$

And the coefficients B_1 and B_2 are identical to A_1 and A_2 of the transformation in Eqs. (7.6) and (7.7).   We shall use A_1 and A_2 in further developments.

Interesting properties of the zero-lift pressure distribution and the additional pressure distribution may be developed as follows:

The lift for a unit chord due to the zero-lift pressure distribution is

$$L_0 = \int_0^{1.0} qP_0 \, dx = 0$$

or *none of the lift results from the zero-lift pressure distribution.*

Now consider the case for the additional pressure distribution:

$$L_a = \int_0^{1.0} c_l P_a q \, dx$$
$$= qc_l = L_{\text{total}}$$

Hence *all of the lift results from the additional pressure distribution.*

The moments yield interesting conclusions, too.   The additional pressure distribution due to lift P_a (it develops) *produces no moment about the quarter chord*, and hence *all of the moment about the quarter chord must be due to the zero-lift pressure distribution P_0.*

If the basic and additional distributions are known, the zero-lift pressure distribution may be readily computed.   Its value, obtained by putting Eq. (7.56) into Eq. (7.55), is

$$P_0 = -\pi A_1 P_a + P_b \qquad (7.61)$$

The value of the coefficient πA_1 in this formula may be determined by noting that the lift coefficient resulting from the basic distribution* alone is

$$c_{lb} = \int_0^{1.0} P_b \, dx$$
$$= 4 \int_0^\pi \sum_1^\infty A_n \sin n\theta \, \tfrac{1}{2} \sin \theta \, d\theta$$

---

* The terminology and symbols used in this section are those of Jacobs in order that reference to his paper may be simplified.   In every case, however, the word *basic* may be replaced by the word *ideal*, and the subscript b by the subscript i.

which, existing only for $n = 1$, becomes

$$c_{lb} = \pi A_1 \qquad (7.62)$$

This equation gives us a simple way to find the important ideal angle, as illustrated in the example below:

Example 7.2. Find the ideal angle and the ideal lift coefficient for the airfoil of Ex. 7.1.

Using Eq. (7.45), the ideal angle is
$$\alpha_I = 623(y_1 - y_5) + 47(y_2 - y_4)$$
$$= 623(0.0001 - 0.0003) + 47(0.0118 - 0.0082)$$
$$= 0.29°$$

Assuming a lift curve slope of 2π and using the angle of zero lift from Ex. 7.1, we get

$$c_{li} = \frac{2}{57.3}(2.13 + 0.29)$$
$$= 0.264$$

or, from Eq. (7.62)

$$c_{li} = c_{lb} = \frac{2\pi(2.13)}{57.3}$$
$$= 0.234$$

The thick-airfoil theory of Chap. 5 gives $c_{li} = 0.282$. The thin-airfoil values would be 0.29 and 0.25 if a thick-airfoil lift curve slope were employed.

Returning to the problem at hand, Eq. (7.61) may be written

$$P_0 = P_b - c_{lb}P_a \qquad (7.63)$$

It now remains to find a method for determining the basic pressure distribution. Equation (7.56) may be written

$$P_b = 4(A_1 \sin \theta + A_2 \sin 2\theta + A_3 \sin 3\theta \cdots A_n \sin n\theta) \qquad (7.64)$$

Since P_0 represents all the moment and P_a none of the moment (about the quarter chord), it is clear from Eq. (7.61) that P_b also represents all the moment. Restating the theorem, all the moment about the quarter chord is due to the pressure distribution at the ideal angle. Hence it seems reasonable to introduce the moment coefficient into Eq. (7.64). From Eq. (7.60)

$$c_{m\frac{1}{4}} = -\frac{\pi}{4}(A_1 - A_2) \qquad (7.65)$$

from which

$$A_1 = -\frac{4}{\pi}c_{m\frac{1}{4}} + A_2$$

and putting this value of A_1 into Eq. (7.64) we get

$$P_b = -\frac{16}{\pi} c_{m\frac{1}{4}} \sin \theta + 4[A_2(\sin \theta + \sin 2\theta) + A_3 \sin 3\theta \cdot \cdot \cdot A_n \sin n\theta]$$

(7.66)

Now all the moment is due to P_b; hence we consider the first term above as representing that portion of the basic pressure distribution due to moment, and thus we write this term in the form

$$-c_{m\frac{1}{4}}P_{bm} = -c_{m\frac{1}{4}}\left(\frac{16}{\pi} \sin \theta\right)$$

and therefore

$$P_{bm} = \frac{16}{\pi} \sin \theta$$

(7.67)

where P_{bm} is the *basic pressure distribution due to moment*.

In the remaining portion of Eq. (7.66) we find the coefficients A_2, A_3, . . . , A_n, all of which are in general proportional to the maximum mean camber-chord ratio h, being part of the series which describes the thin airfoil. We therefore regard this part of the thin airfoil as that portion of the *basic distribution due to camber*, and we write

$$hP_{bc} = 4[A_2(\sin \theta + \sin 2\theta) + A_3 \sin 3\theta \cdot \cdot \cdot A_n \sin n\theta]$$

Hence

$$P_{bc} = \frac{4}{h}[A_2(\sin \theta + \sin 2\theta) + A_3 \sin 3\theta \cdot \cdot \cdot A_n \sin n\theta]$$

(7.68)

The introduction of Eqs. (7.67) and (7.68) in Eq. (7.66) enables us to write the expression for the basic distribution in the form

$$P_b = -c_{m\frac{1}{4}}P_{bm} + hP_{bc}$$

(7.69)

At this point it is in order to discuss the terms *pressure distribution due to camber* and *pressure distribution due to moment* a little more fully since many students find the terms moment and camber very closely related and a certain difficulty arises in differentiating between the action of the two.

For the usual airfoils without reflex (such as the four-digit airfoils, or airfoils having circular arc or parabolic camber) the moment is increased quite directly by increasing the camber [see Eq. (5.52)]. However, two airfoils having the same amount of camber but different shapes of the mean line will in all probability have different moment coefficients, showing that the shape of the mean line is an important parameter. Indeed the moment coefficients of airfoils whose mean lines follow the form of

that shown in Fig. 7.4 are independent of the *amount* of camber. Thus
it is seen that the "moment" and "camber" differentiation is quite
reasonable, *camber* referring to the amount of deviation from a straight
mean line and *moment* to the mean-line shape.

The lift coefficients due, respectively, to these new distributions P_{bm}
and P_{bc} are readily determined by integration, the results being

$$c_{lbm} = \int_0^1 P_{bm}\, dx$$
$$= \int_0^\pi \frac{16}{\pi} \sin\theta \cdot \frac{\sin\theta}{2}\, d\theta = 4.0 \tag{7.70}$$

and

$$c_{lbc} = \frac{4}{h} \int_0^\pi [A_2(\sin\theta + \sin 2\theta) + A_3 \sin 3\theta + \cdots A_n \sin n\theta] \frac{\sin\theta}{2}\, d\theta$$

The above integral exists only for the A_2 term, when it equals $\pi/4$, and
hence

$$c_{lbc} = \frac{\pi A_2}{h} \tag{7.71}$$

The theoretical value of the lift for the entire basic distribution is therefore, from Eq. (7.69),

$$P_b = -c_{m\frac{1}{4}}P_{bm} + hP_{bc}$$
$$c_{lb} = -c_{m\frac{1}{4}}c_{lbm} + hc_{lbc} \tag{7.72}$$
$$c_{lb} = -4c_{m\frac{1}{4}} + \pi A_2 \tag{7.73}$$

Putting in the coefficient value for $c_{m\frac{1}{4}}$ from Eq. (7.65), we get

$$c_{lb} = -4\left[-\frac{\pi}{4}(A_1 - A_2) \right] + \pi A_2 = \pi A_1 \tag{7.74}$$

This corroborates Eq. (7.62) and in addition enables us to follow the
ideal-angle-of-attack concept discussed in Chap. 6 and earlier in this
chapter.

In Eq. (7.54) it appears that one may add an additional pressure distribution to that at zero lift in order to find the pressure distribution at
some angle α, instead of adding the additional to that at the ideal angle.
To make the picture clearer, write the total pressure coefficient as the
sum of the ideal (basic) and the flat-plate (additional) values. Thus

$$P = P_b + (c_l - c_{li})P_a$$

Now, upon substituting from Eq. (7.63) and clearing, we get

$$P = P_0 + c_l P_a$$

showing that, in accordance with the discussion at the end of Chap. 6, this section and the Abbott-von Doenhoff-Stivers theory to follow are both based on the same fundamental assumptions.

For the purpose of introducing empirical corrections that bring the theoretical pressure distribution closer to that realized in practice, Jacobs in Ref. 7.3 writes Eq. (7.69) in the general form as follows:

$$c_{lb} = -c_{m\frac{1}{4}}c_{lbm} + hc_{lbc} \qquad (7.75)$$

The first step is to modify the additional pressure distribution of Eq. (7.53), which, because of the infinitesimally thin mean camber line and

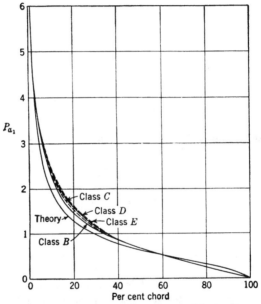

FIG. 7.8.   Comparison of different types of P_{a1} loadings.

zero radius of curvature at the nose, yields an infinitely low pressure at the leading edge ($\theta = 0°$). Further, the growth of the boundary layer in the actual case produces an effective loss in camber which tends to reduce the experimental P_a values below those theoretically derived. The latter effect is more fully discussed in Sect. 8.2. Thus P_a as treated by Jacobs in Ref. 7.3 is partially replaced by P_{a1} as given in Table 7.4 and shown in Fig. 7.8.

Four classes of P_{a1} distributions are given, differing mainly in the values of the peak pressures near the leading edge. Apparently this classification is based on the consideration of the leading-edge radius of the actual airfoil sections, Class B corresponding to a small radius and

Classes C, D, and E to progressively larger ones. The larger the leading-edge radius, the lower will be the peak velocity and pressure in this region.

All of the actual loadings will, of course, go to zero at the leading and trailing edges since there cannot be an infinitely rapid increase of load in real air. Later, for finite wings, we shall apply the same logic to conditions at the wingtips.

A further modification of the additional pressure distribution is made by adding to P_{a1} a term which accounts for the effect of the boundary layer and brings the aerodynamic center into agreement with its experimentally determined position. This correction is noted by the term

$$x_{ac}P_{ac} \tag{7.76}$$

in which x_{ac} is the distance which the actual aerodynamic center is ahead of the quarter-chord point. Values of P_{ac} have already been given in Table 7.4. Thus the corrected additional distribution is given by

$$P_a = P_{a1} + x_{ac}P_{ac} \tag{7.77}$$

The modifications of the basic-moment and basic-camber distributions are not clearly understood. Equations (7.69) and (7.72) are used in the form employed in the theoretical development. Only one class of values for P_{bm} is indicated. The basic-camber distribution P_{bc} involves coefficients which are dependent on the shape of the mean camber line, but the nature of this dependence in the empirical form is not indicated, and the significance of the three classes given for P_{bc} is not known at present. These values are given in Table 7.5 and Fig. 7.9.

Both the additional and basic distribution factors given in Tables 7.4 and 7.5 are classified on the basis of the nature of the airfoil section shape and some of its aerodynamic properties. A classification symbol consisting of a letter and two digits for each airfoil is employed in which the letter indicates the type of P_{a1} distribution, the first number the type of P_{bm} distribution, and the second number the type of P_{bc} distribution. Values of these classification symbols for a number of the more commonly used airfoils are shown in Table 7.6 along with x_{ac} values locating the aerodynamic-center position ahead of the quarter chord, and finally values of the maximum mean camber h_{max} in per cent chord.

The use of the theory and tables is illustrated in the following example:

Example 7.3. Find and plot the normal-force pressure distribution P for an NACA 4412 airfoil at $c_l = 0.4$. (In order to save space the calculations will be shown only for the 5 per cent chord station.)
From Table 7.6 the pressure-distribution class for the NACA 4412 is C 10; $\alpha_{ZL} = -4.0°$, $c_{mac} = -0.088$, $a_0 = 0.098$, and the aerodynamic center is 0.8 per

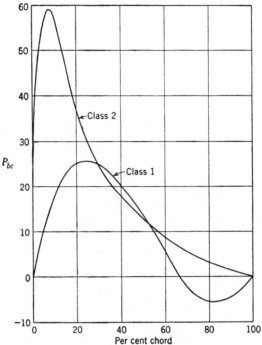

FIG. 7.9.   Comparison of basic camber distributions for Class 1 and 2 airfoils.

cent chord ahead ($x_{ac} = 0.008$) and 2.0 per cent above ($h = 0.02$) the quarter-chord point on the chord line.   From Table 7.4, $P_{ac} = 5.5$, and $P_{a1} = 3.22$ at the 5 per cent chord station.   From Table 7.6, $P_{bm} = 6.05$, $c_{lbm} = 6.30$, $c_{lbc} = 0.00$. and $P_{bc} = 0.0$ at the 5 per cent chord station.

1. $x_{ac}P_{ac} = 0.044$

2. $c_{mac}P_{bm} = 0.532$

3. $hP_{bc} = 0.0$

4. $-c_{mac}c_{lbm} = 0.5544$

5. $hc_{lbc} = 0.0$

6. $c_{lb} = -c_{mac}c_{lbm} + hc_{lbc}$
   $= 0.5544$

7. $P_a = P_{a1} + x_{ac}P_{ac} = 3.264$

8. $P_b = -c_{mac}P_{bm} + hP_{bc} = 0.532$

9. $c_{lb}P_a = 1.809$

10. $P_0 = P_b - c_{lb}P_a = -1.277$

11. $c_lP_a = 1.306$

12. $P = P_0 + c_lP_a = 0.029$

The complete plot is shown in Fig. 7.10.

7.9. Upper- and Lower-surface Pressure Distribution.   Now although the method outlined yields the chordwise *load* distribution, no cognizance has been taken of the specific pressures on upper and lower surfaces. Allen (Ref. 7.4) has outlined a method that furnishes this information in a short and simple manner.

TABLE 7.5. BASIC CHORD LOAD DISTRIBUTION FACTORS FOR AIRFOIL SECTIONS

Sta. per cent chord	P_{bm} Class 1	P_{bc}		
		Class 0	Class 1	Class 2
0	0	0	0	0
1.25	2.85	0	2.5	32.5
2.5	4.25	0	5.5	47.0
5.0	6.05	0	10.0	56.5
7.5	7.10	0	14.5	59.0
10.0	7.80	0	18.0	57.5
15.0	8.80	0	23.0	47.5
20.0	9.30	0	25.0	37.0
30.0	9.50	0	25.0	24.5
40.0	8.80	0	20.5	18.0
50.0	7.75	0	14.0	13.0
60.0	6.60	0	6.0	9.0
70.0	5.30	0	-2.5	5.5
80.0	3.75	0	-5.5	3.5
90.0	2.05	9	-4.5	1.5
95.0	1.10	0	-2.5	1.0
100.0	0	0	0	0
	c_{lbm}	c_{lbc}		
	6.30	0	9.70	18.75

Assume that the curvature of the mean camber line is slight and that the bound vortices that reproduce the thin-airfoil flow pattern (in accordance with our second approach) are located on the mean line as previously considered. Then the induced velocities on the upper and lower surfaces at any given distance x behind the airfoil leading edge are equal in magnitude but opposite in sign. We may write these velocities as

$$u_U = u' + \Delta u \qquad (7.78)$$
$$u_L = u' - \Delta u \qquad (7.79)$$

where Δu = induced velocity due to camber

u' = velocity that the thick no-camber base profile would experience under the same stream conditions

u_U = velocity on upper surface at point x

u_L = velocity on lower surface at point x

If Δp_f = surface pressure increment at point x, we may write for the symmetrical shape

$$p_0 + \tfrac{1}{2}\rho V^2 = p_x + \tfrac{1}{2}\rho(u')^2$$

where V and p_0 are the free-stream velocity and pressure, respectively. Hence

$$\Delta p_f = p_x - p_0 = \tfrac{1}{2}\rho V^2 - \tfrac{1}{2}\rho(u')^2$$

Defining

$$q_0 = \frac{1}{2}\rho V^2 \quad \text{and} \quad P_f = \frac{\Delta p_f}{q_0}$$

we get

$$u' = V \sqrt{1 - P_f} \tag{7.80}$$

TABLE 7.6. AIRFOIL SECTION CHARACTERISTICS

Airfoil	PD classification	Fundamental section characteristics					
		Per cent camber	α_{ZL}	c_{mac}	x_{ac}	y_{ac}	a_0
Clark Y.	C 10	3.9	−5.0	−0.069	1.1	4	0.092
Clark YM-15	D 10	4.0	−5.2	−0.068	1.1	7	0.094
Clark YM-18	E 10	4.0	−5.1	−0.064	1.4	5	0.091
Curtiss C-72	C 10	4.0	−5.6	−0.084	1.0	3	0.095
Göttingen 387	D 10	5.9	−6.6	−0.093	0.7	4	—
Göttingen 398	D 10	4.9	−6.0	−0.081	0.4	1	0.094
N-22	C 10	4.3	−5.4	−0.075	0.6	4	0.096
NACA CYH	C 11	3.1	−2.9	−0.027	0.7	6	0.095
NACA-M6	C 11	2.4	−0.8	−0.002	−0.4	0	0.095
RAF 15	A 10	2.6	−2.2	−0.053	1.7	10	—
USA 27	C 10	5.6	−4.7	−0.078	1.8	5	—
USA 35-A	E 10	7.3	−8.0	−0.111	0.8	5	—
USA 35-B	C 10	4.6	−5.2	−0.076	0.5	5	—
NACA 0006	A 10	0	0	0	0.7	2	0.098
NACA 0012	C 10	0	0	0	0.6	3	0.099
NACA 2212	C 12	2.0	−1.8	−0.029	0.9	5	0.099
NACA 2409	B 10	2.0	−1.7	−0.044	0.7	4	0.099
NACA 2412	C 10	2.0	−2.0	−0.043	0.5	3	0.098
NACA 2415	D 10	2.0	−1.7	−0.040	1.4	5	0.097
NACA 2418	E 10	2.0	−1.9	−0.038	1.1	2	0.094
NACA 4412	C 10	4.0	−4.0	−0.088	0.8	2	0.098
NACA 23006	A 12	1.8	−1.2	−0.012	1.0	8	0.100
NACA 23009	B 12	1.8	−1.1	−0.009	0.9	7	0.099
NACA 23012	C 12	1.8	−1.2	−0.008	1.2	7	0.100
NACA 23015	D 12	1.8	−1.1	−0.008	1.1	6	0.098
NACA 23018	E 12	1.8	−1.2	−0.006	1.7	6	0.097

From Eqs. (7.78) to (7.80)

$$u_U = V \sqrt{1 - P_f} + \Delta u$$
$$u_L = V \sqrt{1 - P_f} - \Delta u$$

The pressure increments on the upper and lower surfaces of the airfoil at x are then

$$\Delta p_U = q_0 - \frac{\rho}{2} (V \sqrt{1 - P_f} + \Delta u)^2 \qquad (7.81)$$

$$\Delta p_L = q_0 - \frac{\rho}{2} (V \sqrt{1 - P_f} + \Delta u)^2 \qquad (7.82)$$

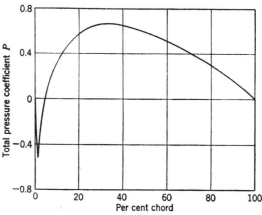

FIG. 7.10.   Normal force total load distribution on an NACA 4412 airfoil at $c_l = 0.4$ by Jacobs' method.

Now if Δp is the normal pressure at station x, being positive upward, then

$$\Delta p = -p_U + p_L \qquad (7.83)$$

Using Eqs. (7.81) to (7.83), we then get

$$\Delta p = 2\rho V \Delta u \sqrt{1 - P_f} \qquad (7.84)$$

from which

$$\Delta u = \frac{1}{4} \frac{\Delta p}{q_0} \frac{V}{\sqrt{1 - P_f}} \qquad (7.85)$$

Putting Δu from Eq. (7.85) into Eqs. (7.81) and (7.82) and defining

$$P = \frac{\Delta p}{q_0}, \qquad P_U = \frac{\Delta p_U}{q_0}, \qquad P_L = \frac{\Delta p_L}{q_0} \qquad (7.86)$$

we get

$$P_U = 1 - \frac{(1 - P_f + \frac{1}{4}P)^2}{1 - P_f} \tag{7.87}$$

and, similarly,

$$P_L = 1 - \frac{(1 - P_f - \frac{1}{4}P)^2}{1 - P_f} \tag{7.88}$$

Hence, if the normal force distribution P over an airfoil is known, and the pressure distribution over the thick no-camber profile is known, then Eqs. (7.87) and (7.88) may be used to get the normal pressure distribution over the upper and lower surfaces. For convenience, a table of $1 - P_f$ for the NACA four-digit and five-digit airfoils is presented. It will be noted there that no values are given for the leading edge $x = 0$. This is because the slope of the airfoil is then infinite and the solution

TABLE 7.7. VALUES OF $1 - P_f$ FOR NACA FOUR- AND FIVE-DIGIT AIRFOIL SECTIONS

Sta. per cent chord	Thickness per cent chord						
	6	9	12*	15	18	21	25
1.25	1.118	1.075	1.008	0.932	0.857	0.775	0.699
2.5	1.190	1.232	1.257	1.250	1.216	1.170	1.105
5.0	1.219	1.313	1.395	1.460	1.505	1.529	1.543
7.5	1.222	1.330	1.425	1.513	1.601	1.675	1.760
10	1.220	1.330	1.432	1.534	1.635	1.737	1.861
15	1.209	1.316	1.428	1.536	1.647	1.760	1.914
20	1.192	1.298	1.407	1.518	1.634	1.749	1.905
25	1.180	1.277	1.380	1.490	1.604	1.715	1.870
30	1.165	1.256	1.352	1.453	1.560	1.667	1.813
35	1.150	1.233	1.321	1.413	1.510	1.603	1.740
40	1.137	1.210	1.290	1.372	1.457	1.540	1.656
45	1.122	1.189	1.256	1.328	1.401	1.473	1.570
50	1.109	1.167	1.225	1.288	1.351	1.410	1.492
55	1.094	1.146	1.195	1.248	1.300	1.350	1.417
60	1.081	1.125	1.167	1.208	1.250	1.293	1.343
65	1.070	1.104	1.138	1.171	1.202	1.239	1.279
70	1.056	1.082	1.107	1.132	1.157	1.182	1.211
75	1.041	1.059	1.074	1.091	1.105	1.122	1.141
80	1.021	1.032	1.041	1.048	1.053	1.059	1.069
85	1.002	1.002	1.001	1.000	0.998	0.995	0.990
90	0.981	0.970	0.955	0.941	0.929	0.911	0.890
95	0.949	0.922	0.895	0.865	0.834	0.807	0.764

* Clark Y; Göttingen 398.

becomes indeterminate. In actual practice it is not difficult to fair the curves in that region. The pressure plots of the airfoils listed in Table 7.6 should be faired to $C_p = 1.0$ at the trailing edge for reasons to be discussed in the next chapter.

Example 7.4. Having the normal-force pressure distribution P for an NACA 4412 at $c_l = 0.4$, separate it into upper- and lower-surface pressures, and plot them.

Considering only the 5 per cent chord point, we have from Example 7.3 the value $P = 0.029$. Then:

1. $\dfrac{P}{4} = 0.0072$

2. $1 - P_f = 1.395$ (from Table 7.6, 12 per cent airfoil)
3. $P_U = -0.405$
4. $P_L = -0.382$

The complete plot is shown in Fig. 7.11.

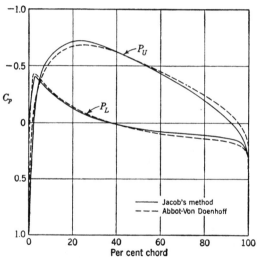

FIG. 7.11. Upper- and lower-surface pressure distribution for NACA 4412 at $c_l = 0.4$, thin-airfoil method. Integrated values $c_n = 0.406$, C. P. $= 46.6$ per cent, $c_{m\frac{1}{4}} = -0.0878$.

7.10. Rapid Estimation of Pressure Distribution. While the previous sections in this chapter have demonstrated the "breakdown" of the behavior of an airfoil, the method as given applies only to airfoils whose P_{a1}, P_{ac}, x_{ac}, P_{bc}, P_{bm}, and $1 - P_f$ values are known. Virtually no hint exists for determining the characteristics of new or modified airfoils.

A complete solution was given by Theodorsen (Ref. 8.1) in 1932. His approach, discussed in the next chapter, solves the arbitrary airfoil problem completely but requires a considerable amount of time and effort.   A second method based on some earlier work by Theodorsen (Ref. 7.5) and completed by Abbott, von Doenhoff, and Stivers (Ref. 7.6) combines the arbitrary airfoil theory with the Munk-Jacobs-Allen tabular approach in a manner that retains a majority of the good features of both.   The method still requires a large number of tables and charts but is much more flexible than the method previously given.

We have mentioned that the pressure distribution over an airfoil may be considered as that due to the pressure distribution at the ideal angle, plus an amount due to angle of attack, referenced to that required for the ideal angle.   In addition, Abbott, von Doenhoff, and Stivers break the ideal distribution into that due to the symmetrical thickness shape and that due to the camber line.   We then have three pressure (and velocity) distributions.

1. The distribution corresponding to the velocity distribution over the basic no-camber thickness form at zero angle of attack v/V.

2. The distribution corresponding to the design load distribution of the mean line $\Delta v/V$, plus on the upper surface.

3. The distribution corresponding to the additional load distribution associated with angle of attack $\Delta v_a/V$.

The pressure coefficient S, defined as

$$S = \frac{H_0 - p}{q_0} = \left(\frac{v}{V} \pm \frac{\Delta v}{V} \pm \frac{\Delta v_a}{V} \right)^2 \tag{7.89}$$

where H_0 = total head in free stream, lb per sq ft

$\quad q_0$ = free-stream dynamic pressure, lb per sq ft

$\quad p$ = pressure on airfoil, lb per sq ft

and, finally,

$$C_p = \frac{\Delta p}{q_0} = 1 - S \tag{7.90}$$

The above addition of velocity increments is not exact theoretically but appears of adequate accuracy for engineering work.

The velocity distributions 1, 2, and 3 above are not fundamentally different* from those of Tables 7.4 to 7.7, but the presentation as velocity increments which may be added and are approximately proportional to their basic parameters gives the newer method much more flexibility. The corrections previously found necessary are incorporated in this new method.   The infinite leading-edge velocity correction (previously in P_{a1}) is in the $\Delta v_a/V$ values since they were calculated by the method of Chap. 8, which properly allows for the leading-edge radius.

---

* Except, of course, that they are referenced to the ideal angle of attack instead of the zero-lift angle.

The use of the tables and charts of Ref. 7.6 is discussed herewith and illustrated by examples below.

1. Values of the basic velocity ratios v/V for airfoils of intermediate thickness ratios may be obtained approximately by linear scaling according to the formula

$$\left(\frac{v}{V}\right)_{t_2} = \left[\left(\frac{v}{V}\right)_{t_1} - 1\right]\frac{t_2}{t_1} + 1 \tag{7.91}$$

where t_1 is a thickness ratio given in the tables.

2. The desired value of $\Delta v/V$ may be found by multiplying the tabular value by the ratio of the desired camber to that of the tabular camber. It is plus for the upper surface, minus for the lower.

3. Values of the additional velocity ratios $\Delta v_a/V$ are given for each chord station. The amount of additional velocity ratio may be found from

$$\frac{\Delta v_a}{V} = \left(\frac{\Delta v_a}{V}\right)_{\text{table}} [c_l - c_{li}] \tag{7.92}$$

where c_l = lift coefficient for which the pressure distribution is desired
c_{li} = ideal lift coefficient

Example 7.5. Using the method of Sect. 7.10, find the pressure coefficient at the 5 per cent chord point of an NACA 4412 at $c_l = 0.4$.

1. From Table 7.8 $\frac{v}{V} = 1.174$ and $\frac{\Delta v_a}{V} = 0.685; \frac{\Delta v}{V} = 0.137$

2. The mean-line data are for 6 per cent camber and $c_{li} = 0.76$. For the desired airfoil (4 per cent camber) we have

$$\frac{\Delta v}{V} = \frac{4}{6}(0.137) = 0.0916 \quad \text{and} \quad c_{li} = \frac{4}{6}(0.76) = 0.506$$

3. Hence $\frac{\Delta v_a}{V} = 0.685\,(0.400 - 0.506)$
$$= -0.0726$$
4. $S_U = (1.174 + 0.0916 - 0.0726)^2$
$$= (1.193)^2 = 1.42$$
5. $\left(\frac{\Delta p}{q}\right)_U = -0.420$
6. $S_L = (1.174 - 0.0916 - 0.0726)^2$
$$= (1.157)^2 = 1.34$$
7. $\left(\frac{\Delta p}{q}\right)_L = -0.34$

These values compare with $(\Delta p/q)_U = -0.405$ and $(\Delta p/q)_L = -0.382$ from Ex. 7.4.

152   BASIC WING AND AIRFOIL THEORY

Integration of the complete pressure diagram indicates a c_l greater than 0.4, and trial and error must be employed with the $\Delta v_a/V$ value [by changing the $(c_l - c_{l_i})$ term] until the proper c_l is found.

While only a single shape and mean-line data table can be given here for reasons of space limitations, Ref. 7.6 contains complete data on 68 thickness shapes and 22 mean-line curves. Since both these parameters may be interpolated for desired proportions, it is seen that the pressure distributions for several thousand airfoils is immediately available.

TABLE 7.8.   VALUES OF v/V AND $\Delta v_a/V$ FOR NACA 0012 BASIC THICKNESS FORM AND $\Delta v/V$ FOR NACA 64 MEAN LINE

x per cent chord	y per cent chord	$\dfrac{v}{V}$	$\dfrac{\Delta v_a}{V}$	$\dfrac{\Delta v}{V}$
0.0	0.0	0.0	1.988	0.0
0.5	—	0.800	1.475	—
1.25	1.894	1.005	1.199	0.064
2.5	2.615	1.114	0.934	0.098
5.0	3.555	1.174	0.685	0.137
7.5	4.200	1.184	0.558	0.167
10.0	4.683	1.188	0.479	0.187
15.0	5.345	1.188	0.381	0.218
20.0	5.737	1.183	0.319	0.242
25.0	5.941	1.174	0.273	0.258
30.0	6.002	1.162	0.239	0.260
40.0	5.803	1.135	0.187	0.250
50.0	5.294	1.108	0.149	0.228
60.0	4.563	1.080	0.118	0.207
70.0	3.664	1.053	0.092	0.188
80.0	2.623	1.022	0.068	0.159
90.0	1.448	0.978	0.044	0.117
95.0	0.807	0.952	0.029	0.084
100.0	0.126	0.0	0.0	0.0

7.11. Summary of Thin-airfoil Theories.   We have seen how the Joukowski transformation has been utilized to determine the angle of zero lift, the ideal angle, and the pitching moment for any airfoil, and that based upon the thin-airfoil theory two methods have been derived that yield the pressure distributions over a large number of airfoils. The first method presented (that of Jacobs and Allen) has been set up for only a limited number of (now) old profiles, while the second method (that of Abbott, von Doenhoff, and Stivers) can handle most of the older airfoils, most of the current NACA profiles, and a great number of others formed from a large selection of mean lines and thickness forms. The second method also requires the use of Ref. 7.6. When a particular airfoil is amenable to both methods, it has been the experience of the author in a limited number of cases that the first method is more accurate.

We need, however, a method that will handle all reasonable airfoils, preferably without any tables or additional charts, and also with a minimum of simplifying assumptions. Such a thick-airfoil theory will be presented in the next chapter.

PROBLEMS

7.1. Show that the additional pressure distribution produces no moment about the quarter chord.

7.2. Following the analysis of Eq. (7.41), show that $\Delta\epsilon_T = 2.92y_1$ if the last 5 per cent of the camber line is parabolic.

7.3. Show that $yf_2(x) = 0$ at $x = 0$ and 1.0.

7.4. Select a cambered airfoil for which test results are available, and calculate and compare α_{ZL} and $c_{m\frac{1}{4}}$ according to the methods in this chapter.

7.5. Place a flat-plate airfoil on the ξ, η axes from $+2a$ to $-2a$, and show how the Joukowski transformation in reverse can transform it into a circle in the z plane.

7.6. Consult Ref. 7.10, and compare the lift increment due to 20 and 40 per cent chord flaps. Assume $c_l = 0.5$ and $\delta_f = 10°$.

7.7. Demonstrate that the center of the z circle is at the origin for the thin-airfoil theory.

7.8. Prove that $\theta = \pi$ corresponds to the trailing edge of the airfoil.

7.9. Show that the thin airfoil has a finite velocity at the trailing edge.

7.10. Prove that the point at which $q_z = 0$ becomes that at which $q_{z'} = 0$.

7.11. Show that $\phi = \pi + \epsilon_T$ is the rear stagnation point of the cylinder.

REFERENCES

7.1. H. Glauert, A Theory of Thin Airfoils, *R & M* 910, 1924.

7.2. Frederick S. Woods, "Advanced Calculus," Ginn & Company, Boston, 1932.

7.3. Eastman N. Jacobs, Airfoil Section Characteristics as Applied to the Prediction of Air Forces and Their Distribution on Wings, *TR* 631, 1938.

7.4. H. Julian Allen, A Simplified Method for the Calculation of Airfoil Pressure Distribution, *TN* 708, 1939.

7.5. T. Theodorsen, On the Theory of Wing Sections with Particular Reference to the Lift Distribution, *TR* 383, 1931.

7.6. Ira H. Abbott, Albert E. von Doenhoff, and Louis S. Stivers, Jr., Summary of Airfoil Data, *TR* 824, 1948.

7.7. M. J. Thompson, Applied Airfoil Theory, unpublished notes, University of Texas, Austin, Tex., 1944.

7.8. H. Glauert, "Aerofoil and Airscrew Theory," Cambridge University Press, London, 1926.

7.9. H. Glauert, Theoretical Relationships for an Airfoil with a Hinged Flap, *R & M* 1095, 1927.

7.10. R. M. Pinkerton, Analytical Determination of the Load on a Trailing Edge Flap, *TN* 353, October, 1930.

7.11. H. Julian Allen, The Calculation of Chordwise Load Distribution over Airfoils Sections with Plain, Split, or Serially Hinged Trailing Edge Flaps, *TR* 634, 1938.

CHAPTER 8

THICK-AIRFOIL THEORIES

8.1. Theodorsen's Theory. Theodorsen (Ref. 8.1) extended the attack of Glauert given in Chap. 7 to include airfoils of finite thickness, finally obtaining a complete solution for the theoretical pressure distribution about an arbitrary airfoil at reasonable angles of attack.

As we shall see, his attack parallels that of Glauert in that he used the Joukowski transformation "in reverse" and then devised a general trans-

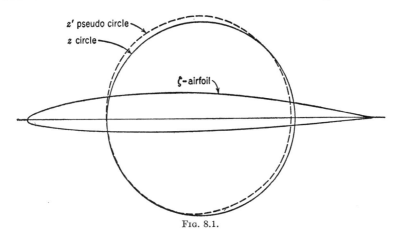

Fig. 8.1.

formation to get from the pseudo circle to the circle. This time, by means of a different (and more complex) procedure, it is not necessary to reduce the airfoil to its mean line.

The airfoil is described (to use Theodorsen's notation) by

$$\zeta = x + iy$$

while the circle is

$$z = ae^{\psi_0}e^{i\varphi}$$

where a is a constant and ψ_0 is a small constant so that e^{ψ_0} nearly equals 1.0. The pseudo circle (see Fig. 8.1)

$$z' = ae^{\psi}e^{i\theta}$$

155

has the term e^{ψ} [where ψ is a $f(\theta)$] to vary the radius and produce the irregular curve. Points on the circle are out of angular agreement with points of the pseudo circle by $\epsilon = \varphi - \theta$.

It has already been noted that the circulation will be unchanged by the transformations.

We first transform the airfoil into the pseudo circle by

$$\zeta = z' + \frac{a^2}{z'} \tag{8.1}$$

We previously have noted that at great distances (large z') $\zeta \cong z'$, showing that the flow at great distances is similar in both planes. Near the origin the flows are entirely different but related by Eq. (8.1).

Let us now expand Eq. (8.1) as

$$\zeta = ae^{\psi+i\theta} + ae^{-\psi-i\theta}$$

and hence

$$\zeta = a(e^{\psi} + e^{-\psi}) \cos \theta + ia(e^{\psi} - e^{-\psi}) \sin \theta \tag{8.2}$$

Recalling the basic hyperbolic functions [Eq. (2.31)], we have from Eq. (8.2)

$$\zeta = 2a \cosh \psi \cos \theta + 2ia \sinh \psi \sin \theta$$

Since

$$\zeta = x + iy$$
$$x = 2a \cosh \psi \cos \theta$$

and

$$y = 2a \sinh \psi \sin \theta$$

from which

$$\cosh \psi = \frac{x}{2a \cos \theta} \tag{8.3}$$

$$\sinh \psi = \frac{y}{2a \sin \theta}$$

From the fundamental relation that

$$\cosh^2 \psi - \sinh^2 \psi = 1$$

we get

$$\frac{x^2}{4a^2 \cos^2 \theta} - \frac{y^2}{4a^2 \sin^2 \theta} - 1 = 0 \tag{8.4}$$

from which by clearing and factoring and defining

$$p = 1 - \frac{x^2}{4a^2} - \frac{y^2}{4a^2} \tag{8.5}$$

we get

$$4 \sin^4 \theta - 4p \sin^2 \theta + p^2 = p^2 + \left(\frac{y}{a}\right)^2$$

and

$$2 \sin^2 \theta - p = \sqrt{p^2 + \left(\frac{y}{a}\right)^2} \tag{8.6}$$

Equation (8.6) relates the argument θ of the pseudo circle with the coordinates of the airfoil x and y.

We can similarly solve for the modulus factor ψ, getting

$$2 \sinh^2 \psi = -p + \sqrt{p^2 + \left(\frac{y}{a}\right)^2}$$

but it is more convenient to express ψ as a series since it is usually small and the process simpler thereby. We proceed as follows:

From Eq. (2.31) we may write

$$e^\psi = \sinh \psi + \cosh \psi$$
$$= \sinh \psi + \sqrt{1 + \sinh^2 \psi}$$

For $\sinh \psi$ small we may expand the radical as a binomial, getting

$$e^\psi = \sinh \psi + 1 + \tfrac{1}{2} \sinh^2 \psi + \cdots$$

so that

$$\psi = \ln e^\psi = \ln (1 + \sinh \psi + \tfrac{1}{2} \sinh^2 \psi + \cdots)$$

Now we may write (for x small)

$$\ln (1 + x) = x - \frac{x^2}{2} + \frac{x^3}{3} - \frac{x^4}{4} + \cdots$$

Thus

$$\psi = \ln [1 + (\sinh \psi + \tfrac{1}{2} \sinh^2 \psi)]$$
$$= \sinh \psi - \tfrac{1}{6} \sinh^3 \psi$$

From Eq. (8.3) this becomes

$$\psi = \frac{y}{2a \sin \theta} - \frac{1}{6} \left(\frac{y}{2a \sin \theta}\right)^3 \tag{8.7}$$

as long as $\psi < \ln 2$.

We may now reproduce the conformal representation in the z' plane since for each point of the airfoil x, y (and hence p) both θ and ψ have been determined by Eqs. (8.6) and (8.7). However, the points at the leading edges offer some difficulty as both θ and y are zero and ψ becomes indeterminate. We approach this problem by rewriting Eq. (8.4) as

$$\left(\frac{x}{2a \cosh \psi}\right)^2 + \left(\frac{y}{2a \sinh \psi}\right)^2 = 1.0 \tag{8.8}$$

If for a minute we imagine ψ constant, we see that Eq. (8.8) is that of an ellipse, the variation of ψ at each point in the actual case changing it into an airfoil. Obviously, it is of advantage to make ψ as small as possible, and this we accomplish by making the airfoil as elliptic as possible, that is, by placing it on axes so that it most nearly approximates an ellipse. To do this, we note that the x and y intercepts of Eq. (8.8) are at $\pm 2a \cosh \psi$ and $\pm 2a \sinh \psi$, and the foci at

$$\sqrt{4a^2 \left(\cosh^2 \psi - \sinh^2 \psi\right)} = \pm 2a,\ 0$$

Solving Eq. (8.8) for y, we get

$$y = 2a \sinh \psi \sqrt{1 - \left(\frac{x}{2a \cosh \psi}\right)^2} \tag{8.9}$$

from which

$$\frac{dy}{dx} = \frac{(-2a \sinh \psi)x}{(2a \cosh \psi)^2 \sqrt{1 - \left(\frac{x}{2a \cosh \psi}\right)^2}}$$

and

$$\frac{d^2y}{dx^2} = \frac{(-2a \sinh \psi)(2a \cosh \psi)\left[1 - \left(\frac{x}{2a \cosh \psi}\right)^2\right] - (2a \sinh \psi)x}{(2a \cosh \psi)^4 \left(1 - \frac{x}{2a \cosh \psi}\right)^{\frac{3}{2}}}$$

Recalling that the radius of curvature of a curve is given by

$$\rho = \frac{[1 + (dy/dx)^2]^{\frac{3}{2}}}{d^2y/dx^2}$$

and that the airfoil leading edge is at $x_N = 2a \cosh \psi_N$, we get

$$\rho_N = \frac{(2a \sinh \psi_N)^2}{2a \cosh \psi_N} = 2a\psi_N^2 \tag{8.10}$$

since ψ_N is small.

Finally

$$\psi_N = \sqrt{\frac{\rho_N}{2a}} \tag{8.11}$$

Putting $\cosh \psi = (1 + \sinh^2 \psi)^{\frac{1}{2}}$ and expanding by the binomial theorem we get

$$x_0 = 2a \left(1 + \frac{\psi_N{}^2}{2}\right)$$

$$= 2a + \frac{\rho_N}{2} \tag{8.12}$$

Hence the length $4a$ corresponds to the distance from a point midway between the nose and the center of curvature of the leading edge, to a point midway between the tail and the center of curvature of the trailing edge. Normally the trailing edge is sharp so that the first step in the Theodorsen process is to determine a by taking one-quarter of the airfoil less one-half the leading-edge radius (see Fig 8.2).

To find the velocity at a point x, y on the airfoil, we start in the customary manner with the velocity around a circle in two-dimensional

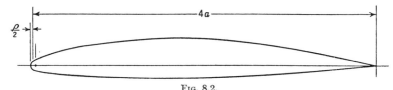

Fɪɢ. 8.2.

flow. Contrary to our previous practice, however, we make the radius of the circle ae^{ψ_0} where ψ_0 is a small constant quantity which will later be shown to represent the average value of ψ taken around the z circle.

The potential function of the flow past the circle is [from Eq. (4.51)]

$$w = -V \left(z + \frac{a^2 e^{2\psi_0}}{z}\right) - i \frac{\Gamma}{2\pi} \ln \frac{z}{ae^{\psi_0}} \tag{8.13}$$

The velocity

$$\frac{dw}{dz} = -V \left(1 - \frac{a^2 e^{2\psi_0}}{z^2}\right) - \frac{i\Gamma}{2\pi z} \tag{8.14}$$

Now the velocity must vanish at the point on the circle which later becomes the airfoil trailing edge [see Eq. (5.14)], and so we may write

$$z_T = -ae^{\psi_0}e^{i(\alpha + \epsilon_T)} \tag{8.15}$$

We have then

$$\frac{dw}{dz} = 0$$

and

$$\Gamma = \frac{-2\pi z_T V}{i} \left(1 - \frac{a^2 e^{2\psi_0}}{z_T{}^2}\right) \tag{8.16}$$

From Eqs. (8.15) and (2.34) the circulation is

$$\Gamma = 4\pi ae^{\psi_0} V \sin(\alpha + \epsilon_T) \tag{8.17}$$

This flow may now be transformed into the flow around the pseudo circle. As mentioned previously the circulation will be unaltered, and $L = \rho V\Gamma$ as usual. We seek to transform the circle $z = ae^{\psi_0 + i\varphi}$ into the pseudo circle $z' = ae^{\psi + i\theta}$. Let us consider a general transformation that will fit the case in hand. We shall need a general series for the variation of the modulus, say $e^{\sum_1^\infty A_n}$; one for the argument, say $e^{\sum_1^\infty iB_n}$; and a $\dfrac{1}{z^n}$ term to assure that the flow is unaltered at infinity.

Hence

$$z' = ze^{\sum_1^\infty (A_n + iB_n)\frac{1}{z^n}} \tag{8.18}$$

Now by our definitions

$$\frac{z'}{z} = \frac{ae^{(\psi + i\theta)}}{ae^{(\psi_0 + i\varphi)}}$$

so that

$$z' = ze^{(\psi - \psi_0) + i(\theta - \varphi)} \tag{8.19}$$

From Eqs. (8.18) and (8.19)

$$\psi - \psi_0 + i(\theta - \varphi) = \sum_1^\infty (A_n + iB_n)\frac{1}{z^n} \tag{8.20}$$

Now a general point P in the z plane may be expressed in polar form

$$z = re^{i\varphi} = r(\cos\varphi + i\sin\varphi)$$

($z = ae^{\psi_0}e^{i\varphi}$ for the circle only), and by de Moivre's theorem

$$z^{-n} = r^{-n}e^{-in\varphi}$$

so that

$$\frac{1}{z^n} = \frac{1}{r^n}(\cos n\varphi - i\sin n\varphi)$$

and Eq. (8.20) becomes

$$\psi - \psi_0 + i(\theta - \varphi) = \sum_1^\infty (A_n + iB_n)\frac{1}{r^n}(\cos n\varphi - i\sin n\varphi)$$

Equating reals and imaginaries,

$$\psi - \psi_0 = \sum_1^\infty \left(\frac{A_n}{r^n}\cos n\varphi + \frac{B_n}{r^n}\sin n\varphi\right) \tag{8.21}$$

and

$$\theta - \varphi = \sum_1^\infty \left(\frac{B_n}{r^n} \cos n\varphi - \frac{A_n}{r^n} \sin n\varphi \right) \tag{8.22}$$

The coefficients A_n/r^n, B_n/r^n as well as ψ_0 may be determined from Eq. (8.21) as follows: From Eq. (8.21)

$$\psi = \psi_0 + \sum_1^\infty \left(\frac{A_n}{r^n} \cos n\varphi + \frac{B_n}{r^n} \sin n\varphi \right)$$

This is a Fourier series of the form

$$f(x) = \tfrac{1}{2}a_0 + \sum_1^\infty (a_n \cos nx + b_n \sin nx)$$

From page 458 of Ref. 8.3 the solution is simply

$$a_n = \frac{1}{\pi} \int_0^{2\pi} f(x) \cos nx \, dx$$

$$b_n = \frac{1}{\pi} \int_0^{2\pi} f(x) \sin nx \, dx$$

and

$$a_0 = \frac{1}{\pi} \int_0^{2\pi} f(x) \, dx$$

which becomes in this case

$$a_n = \frac{A_n}{r^n} = \frac{1}{\pi} \int_0^{2\pi} \psi \cos n\varphi \, d\varphi \tag{8.23}$$

$$b_n = \frac{B_n}{r^n} = \frac{1}{\pi} \int_0^{2\pi} \psi \sin n\varphi \, d\varphi \tag{8.24}$$

$$a_0 = 2\psi_0 = \frac{1}{\pi} \int_0^{2\pi} \psi \, d\varphi \tag{8.25}$$

Recalling that φ is the argument of z, we see that ψ_0 is the average value of ψ taken around the circle.

Since the quantity $\theta - \varphi$ is needed in the following analysis, we eliminate the coefficients in Eq. (8.22) by using Eqs. (8.23) and (8.24). A subscript c will be attached to those angles which are held constant while the integration to evaluate the coefficients A_n/r^n, B_n/r^n is being performed.

Hence

$$(\theta - \varphi)_c = \sum_{1}^{\infty} \left(\cos n\varphi_c \frac{1}{\pi} \int_0^{2\pi} \psi \sin n\varphi \, d\varphi - \sin n\varphi_c \frac{1}{\pi} \int_0^{2\pi} \psi \cos n\varphi \, d\varphi \right)$$

The expression may be simplified to

$$(\theta - \varphi)_c = \frac{1}{\pi} \sum_{1}^{\infty} \int_0^{2\pi} \psi \, (\sin n\varphi \cos n\varphi_c - \cos n\varphi \sin n\varphi_c) \, d\varphi$$

$$= \frac{1}{\pi} \sum_{1}^{\infty} \int_0^{2\pi} \psi \sin n(\varphi - \varphi_c) \, d\varphi \qquad (8.26)$$

Now a trigonometric relation exists that shows

$$\sum_{1}^{n} \sin n\alpha = \frac{1}{2} \cot \frac{\alpha}{2} - \frac{\cos (2n + 1)(\alpha/2)}{2 \sin (\alpha/2)} \qquad (8.27)$$

and hence Eq. (8.26) may be written

$$(\theta - \varphi)_c = \frac{1}{2\pi} \int_0^{2\pi} \psi \cot \frac{(\varphi - \varphi_c)}{2} \, d\varphi$$

$$- \frac{1}{2\pi} \int_0^{2\pi} \frac{\psi \cos (2n + 1)[(\varphi - \varphi_c)/2]}{\sin [(\varphi - \varphi_c)/2]} \, d\varphi$$

From page 368 of Ref. 8.3 we see that the second integral is equal to zero, and therefore

$$(\theta - \varphi)_c = \frac{1}{2\pi} \int_0^{2\pi} \psi \cot \frac{\varphi - \varphi_c}{2} \, d\varphi \qquad (8.28)$$

We set $(\theta - \varphi)_c = \epsilon_c$ and find after considerable analysis (see Appendix 2, page 281) that

$$\epsilon_c = -\frac{1}{\pi} [0.628\psi_c' + 1.0647(\psi_1 - \psi_{-1}) + 0.4431(\psi_2 - \psi_{-2})$$

$$+ 0.2311(\psi_3 - \psi_{-3}) + 0.1030(\psi_4 - \psi_{-4})] \qquad (8.29)$$

where $\psi_c' = $ slope of ψ curve at $\varphi = \varphi_c$

$\psi_1 = $ value of ψ at $\varphi + \frac{\pi}{5}$, ψ_{-1} at $\varphi - \frac{\pi}{5}$, etc.

We now resume the task of determining the velocity at any point on the surface of the airfoil. The velocity at the surface of the circle is dw/dz. For corresponding points in the z' plane and on the airfoil itself we have $\dfrac{dw}{dz} \cdot \dfrac{dz}{dz'}$ and $\dfrac{dw}{dz} \cdot \dfrac{dz}{dz'} \cdot \dfrac{dz'}{d\zeta}$. The quantities ζ and z are related by the Joukowski transformation, which yields upon differentiating,

$$\frac{d\zeta}{dz'} = \frac{1}{z'}\left(z' - \frac{a^2}{z'}\right)$$
$$= \frac{1}{z'}(ae^{\psi+i\theta} - ae^{-\psi-i\theta})$$

Expanding, we get

$$\frac{d\zeta}{dz'} = \frac{1}{z'}(2a \sinh \psi \cos \theta + 2ia \cosh \psi \sin \theta)$$

and from Eq. (8.3) we then have

$$\frac{d\zeta}{dz'} = \frac{1}{z'}(y \cot \theta + ix \tan \theta) \tag{8.30}$$

It now remains to find $\dfrac{dz}{dz'}$. From (Eq. 8.18)

$$\frac{dz'}{dz} = z\left[e^{\sum_1^\infty (A_n+iB_n)\frac{1}{z^n}} \frac{d}{dz}\left(\sum_1^\infty A_n + iB_n\right)\frac{1}{z^n}\right] + e^{\sum_1^\infty (A_n+iB_n)\frac{1}{z^n}}$$
$$= z' \cdot \frac{d}{dz}\left[\sum_1^\infty (A_n + iB_n)\frac{1}{z^n}\right] + \frac{z'}{z}$$

and from Eq. (8.20)

$$\frac{dz'}{dz} = z'\left\{\frac{d}{dz}[(\psi - \psi_0) + i(\theta - \varphi)] + \frac{1}{z}\right\}$$
$$= z'\left\{\frac{d}{dz}[(\psi - \psi_0) + i(\theta - \varphi) + \ln z]\right\} \tag{8.31}$$

Since

$$\ln z = \ln(ae^{\psi_0} \cdot e^{i\varphi}) = \ln a + \psi_0 + i\varphi$$

Eq. (8.31) becomes

$$\frac{dz'}{dz} = z'\frac{d}{dz}(\psi - \psi_0 + i\theta - i\varphi + \ln a + \psi_0 + i\varphi)$$
$$= z'\frac{d}{d\theta}(\psi + i\theta) \cdot \frac{d\theta}{dz} \tag{8.32}$$

Now

$$z = ae^{\psi_0}e^{i\varphi}$$

or

$$dz = ae^{\psi_0}e^{i\varphi} \cdot i\,d\varphi = iz\,d\varphi$$

and

$$\frac{dz}{z} = i \, d\varphi = i \, d(\varphi - \theta) + i \, d\theta$$

from which

$$\frac{dz}{d\theta} = iz \left[1 + \frac{d(\varphi - \theta)}{d\theta} \right]$$

So from Eq. (8.32)

$$\frac{dz'}{dz} = z' \frac{d}{d\theta} \left(\psi + i\theta\right) \cdot \frac{1}{iz \left[1 + \dfrac{d(\varphi - \theta)}{d\theta} \right]}$$

If we put $\epsilon = \varphi - \theta$ and call the slopes of ϵ and ψ against θ, ϵ' and ψ', respectively, we get

$$\frac{dz'}{dz} = \frac{z'}{z} \frac{1 - i\psi'}{1 + \epsilon'} \tag{8.33}$$

Equations (8.30) and (8.33) may be combined as

$$\frac{d\zeta}{dz} = \frac{d\zeta}{dz'} \frac{dz'}{dz} = \frac{1}{z} \left(y \cot \theta + ix \tan \theta\right) \frac{1 - i\psi'}{1 + \epsilon'}$$

Expanding and collecting reals and imaginaries, we have

$$\frac{d\zeta}{dz} = \frac{(y \cot \theta + x\psi' \tan \theta) + i(x \tan \theta - y\psi' \cot \theta)}{z(1 + \epsilon')}$$

Now, as we are interested rather in the magnitude of the velocity than in its direction, we take the square root of the sum of the squares of the reals and imaginaries. After some algebra we have

$$\left|\frac{d\zeta}{dz}\right| = \frac{\sqrt{(x^2 \tan^2 \theta + y^2 \cot^2 \theta)[1 + (\psi')^2]}}{ae^{(\psi_0 + i\varphi)} \cdot (1 + \epsilon')}$$

Bringing the factor a up under the radical, we have

$$\left|\frac{d\zeta}{dz}\right| = \frac{2\sqrt{\left[\left(\dfrac{y}{2a \sin \theta}\right)^2 \cos^2 \theta + \left(\dfrac{x}{2a \cos \theta}\right)^2 \sin^2 \theta\right][1 + (\psi')^2]}}{e^{\psi_0}(1 + \epsilon')}$$

The factor $e^{i\varphi}$ is directional and is of no importance in considering the absolute value of the velocity. After using Eqs. (8.3) and (8.4) we have finally

$$\left|\frac{d\zeta}{dz}\right| = \frac{2\sqrt{\left[\left(\dfrac{y}{2a \sin \theta}\right)^2 + \sin^2 \theta\right][1 + (\psi')^2]}}{e^{\psi_0}(1 + \epsilon')} \tag{8.34}$$

The numerical value of the velocity at the surface of the circle is obtained from Eqs. (8.14) and (8.17) as follows:

Substitute the general point

$$z = ae^{\psi_0 + i(\alpha + \varphi)}$$

where α is the angle of attack as measured from the axis coordinates in Eq. (8.14). We get

$$\frac{dw}{dz} = -V\left[1 - \frac{a^2 e^{2\psi_0}}{a^2 e^{2\psi_0} \cdot e^{2i(\alpha + \varphi)}}\right] - i\frac{4\pi a V e^{\psi_0} \sin(\alpha + \epsilon_T)}{2\pi a e^{\psi_0} e^{i(\alpha + \varphi)}}$$

which cancels and expands to

$$\frac{dw}{dz} = -V\{[1 - \cos 2(\alpha + \varphi) + 2\sin(\alpha + \epsilon_T)\sin(\alpha + \varphi)]$$
$$+ i[\sin 2(\alpha + \varphi) + 2\sin(\alpha + \epsilon_T)\cos(\alpha + \varphi)]\}$$

To solve for the absolute value of $\dfrac{dw}{dz}$, we have

$$\left|\frac{dw}{dz}\right|^2 = V^2[4\sin^2(\alpha + \epsilon_T) + 8\sin(\alpha + \epsilon_T)\sin(\alpha + \varphi) + 4\sin^2(\alpha + \varphi)]$$

and

$$\left|\frac{dw}{dz}\right| = 2V[\sin(\alpha + \epsilon_T) + \sin(\alpha + \varphi)] \tag{8.35}$$

Replace φ by $\theta + \epsilon$. It will be recalled that ϵ_T is the value of $\varphi - \theta$ at the tail, and hence, by the Kutta-Joukowski law, it is also the angle of zero lift. We now have from Eqs. (8.35) and (8.34) that the absolute velocity for a point on the airfoil (using v instead of q_{ζ})

$$v = \left|\frac{dw}{dz}\right|\left|\frac{dz}{d\zeta}\right| = \frac{V[\sin(\alpha + \theta + \epsilon) + \sin(\alpha + \epsilon_T)](1 + \epsilon')e^{\psi_0}}{\sqrt{(\sinh^2\psi + \sin^2\theta)[1 + (\psi')^2]}} \tag{8.36}$$

Letting

$$F = \frac{(1 + \epsilon')e^{\psi_0}}{\sqrt{(\sinh^2\psi + \sin^2\theta)[1 + (\psi')^2]}} \tag{8.37}$$

we have

$$v = FV[\sin(\alpha + \theta + \epsilon) + \sin(\alpha + \epsilon_T)]$$

By Eq. (8.17) this becomes

$$v = FV\left[\sin(\alpha + \theta + \epsilon) + \frac{\Gamma}{4\pi a e^{\psi_0} V}\right] \tag{8.38}$$

which is known as the general equation for the local velocity about an airfoil.

The local pressure coefficient may then be found from

$$C_p = 1 - \left(\frac{v}{V}\right)^2 \tag{8.39}$$

A practical application of the Theodorsen theory represents a considerable amount of work—work that must be performed with the utmost care and accuracy. Slide-rule accuracy is definitely insufficient. At least 18 points should be taken on upper and lower surfaces each, and it will take at least 63 columns of calculations, a total of perhaps 60 hours of work. The first 54 columns are for the airfoil itself and the remainder for the specific angle of attack. Hence the pressure distributions for additional angles of attack are much easier to obtain than those of the first one.

To recapitulate, the procedure for calculating the pressure distribution of an arbitrary airfoil should be accomplished as follows:

1. Plot the airfoil to an exaggerated vertical scale in order to check the ordinates x_1, y_1. The Theodorsen theory is accurate and will show up any irregularities.

2. Draw the x axis through the mid-points of the leading-edge and trailing-edge radii, and erect the y axis through the mid-point of the length $4a$, defined by Eq. (8.12).

3. Determine the numerical values of the airfoil ordinates x, y on the new axis. (Not infrequently the x axis is so near the chord that $y_1 = y$.)

4. Determine p from Eq. (8.5) and $\sin^2 \theta$, $\sin \theta$, and θ (in both radians and degrees) from Eq. (8.6).

5. Find ψ from Eq. (8.7), and plot it against θ in radians.

6. Find the slope of the ψ curve $\dfrac{d\psi}{d\theta}$ ($= \psi'$) at each of the points being calculated. This is probably most accurately done by placing a shiny piece of metal on the curve at the point in question and rotating it about a vertical axis until the reflection of the curve appears to smoothly extend the curve itself. Repeat the process from right to left. The average of the two runs will be a good approximation to a true perpendicular to the curve and may be used to get the slope of the curve. Next plot $\dfrac{d\psi}{d\theta}$ vs. θ to aid in evaluating and checking the slope values.

7. Find ψ_0 by integrating the area under the ψ curve and dividing by 2π in accordance with Eq. (8.25).

8. Evaluate ϵ by means of the ψ and ψ' curves and Eq. (8.29). Plot ϵ vs. θ, and find $\dfrac{d\epsilon}{d\theta}$ $(=\epsilon')$ as in step 6. Plot ϵ' vs. θ to aid in checking the accuracy of ϵ'.

9. Find F from Eq. (8.37).

10. For the selected α, Eq. (8.38) can now be used to yield $\dfrac{v}{V}$, and finally C_p is found from Eq. (8.39).

It will be noted by referring to Eqs. (8.25) and (8.29) that they should have been evaluated through the use of curves plotted against φ and not θ, but unfortunately φ cannot be obtained until ϵ is known. Increased accuracy may be obtained if, after step 8 is completed, φ is determined by adding ϵ to θ and a new curve of ψ against φ is drawn. More accurate values from Eqs. (8.25) and (8.29) may then be obtained.

The ideal angle of attack, if desired, may be found from Eq. (8.36) by setting v and θ equal to zero. Calling ϵ at the leading edge ϵ_N, we then get

$$\alpha_I = -\frac{\epsilon_N + \epsilon_T}{2} \tag{8.40}$$

and the ideal angle occurs when the front stagnation point is at the airfoil leading edge.

Conditions at the trailing edge may be investigated by analyzing Eq. (8.36) at that point. Consider, for instance, $\theta = \pi + \Delta\theta$ and $\epsilon = \epsilon_T + \epsilon' \Delta\theta$, where $\Delta\theta$ is very small. Substituting these values into Eq. (8.36) and noting that ψ is always small near the trailing edge gives

and

$$\left|\frac{v_T}{V}\right| = \frac{e^{\psi_0}(1 + \epsilon_T')^2 \Delta\theta}{\sqrt{(\psi_T^2 + \Delta\theta^2)[1 + (\psi_T')^2]}}$$

$$|v_T| = \frac{e^{\psi_0}(1 + \epsilon_T')^2 V}{\sqrt{\left[1 + \left(\dfrac{\psi_T}{\Delta\theta}\right)^2\right][1 + (\psi_T')^2]}}$$

Writing ψ near the trailing edge as

$$\psi_{NT} = \psi_T + \psi_T' \Delta\theta + \tfrac{1}{2}\psi'' \Delta\theta^2$$

we get

$$\frac{\psi_{NT}}{\Delta\theta} = \frac{\psi_T}{\Delta\theta} + \psi_T' + \frac{1}{2}\psi'' \theta\Delta$$

If ψ_T is not zero, then $\psi_T/\Delta\theta$ will be infinite at the trailing edge ($\Delta\theta = 0$) and v will be zero, resulting in a stagnation point at the trailing edge of the airfoil. If, however, the tail is perfectly sharp (*i.e.*, cusped), ψ_T will then be zero and $\psi_T/\Delta\theta = \psi_T'$ for $\Delta\theta = 0$. The velocity at the trailing edge will then be

$$|v_T| = \frac{e^{\psi_0}(1 + \epsilon_T')^2 V}{[1 + (\psi_T')^2]}$$

Normally v_T for cusped airfoils is about $0.8V$ so that $C_p = 0.3$.

The physical significance of this trailing-edge action may perhaps be best seen by noting that normal components exist at the trailing edge of noncusped airfoils which must vanish when the flow becomes parallel past the airfoil. Hence, one may reasonably expect higher trailing-edge pressures for noncusped airfoils.

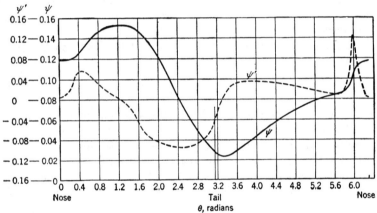

Fig. 8.3. Plot of ψ vs. θ needed to determine ψ' in Ex. 8.1. ψ' is shown as broken line.

Theodorsen in Ref. 8.8 discusses a method of modifying an airfoil through alterations to the ϵ curve and subsequent alterations to the airfoil ordinates. It will be noted that as long as ϵ_N and ϵ_T are held constant other changes to the ϵ curve will have no effect on α_{zL}, c_{li}, and the ideal angle. Indeed, as far as conditions for the ideal angle and lift are concerned it is necessary only to maintain the *difference* between ϵ_N and ϵ_T, in accordance with Eq. (8.40).

The use of the theory in this chapter is exemplified below:

Example 8.1. Calculate the pressure distribution about the modified NACA 65,311 airfoil shown in Fig. 7.5 at $\alpha = 0.5°$.

The results are most simply outlined in tabular form. Explanation of the column headings is on the page following Table 8.1. (For the sake of brevity, the procedure involved in plotting the values against ϕ has not been included.)

TABLE 8.1

1	2	3	4	5	6	7	8	9	10	11	12	13	14
x, per cent chord	y, per cent chord	z_1/c	y_1/c	p	$\sin\theta$	θ, radians	ψ	ψ'	ϵ	ϵ'	F	$\dfrac{v}{V}$	C_p
0	0	−2.0167	0	−0.0167	0.0173	0.0173	(0.1186)	—					
1.25	1.295	−1.9665	0.0520	0.0326	0.2175	0.2193	0.1193	0.0183	−0.0339	0.0077	4.460	1.0628	−0.129
2.50	1.871	−1.9163	0.0752	0.0806	0.3090	0.3141	0.1214	0.0450	−0.0360	0.0138	3.350	1.0998	−0.210
5.00	2.807	−1.8158	0.1127	0.1725	0.4349	0.4499	0.1294	0.0645	−0.0317	0.0149	2.444	1.1340	−0.286
7.50	3.506	−1.7154	0.1408	0.2594	0.5264	0.5545	0.1334	0.0557	−0.0304	0.0300	2.080	1.1507	−0.324
10.00	4.170	−1.6150	0.1705	0.3406	0.6010	0.6445	0.1414	0.0452	−0.0257	0.0385	1.850	1.1701	−0.369
15.00	5.250	−1.4142	0.2109	0.4889	0.7150	0.7963	0.1469	0.0300	−0.0178	0.0480	1.578	1.1893	−0.414
20.00	5.970	−1.2133	0.2398	0.6175	0.7995	0.9264	0.1494	0.0212	−0.0098	0.0600	1.432	1.2088	−0.461
30.00	6.980	−0.8117	0.2804	0.8156	0.9158	1.1571	0.1525	0.0065	0.0011	0.0840	1.283	1.2380	−0.533
40.00	7.480	−0.4100	0.3004	0.9354	0.9793	1.3682	0.1528	−0.0085	0.0220	0.1000	1.220	1.2577	−0.582
50.00	7.410	−0.0083	0.2976	0.9779	1.0000	1.5705	0.1483	−0.0363	−0.0433	0.0910	1.186	1.2333	−0.521
60.00	6.760	0.3933	0.2715	0.9429	0.9809	1.7665	0.1380	−0.0717	−0.0575	0.0690	1.183	1.1959	−0.430
70.00	5.610	0.7950	0.2253	0.8293	0.9189	1.9756	0.1223	−0.0834	−0.0683	0.0410	1.230	1.1456	−0.312
80.00	4.172	1.1967	0.1676	0.6350	0.8039	2.2082	0.1040	−0.0905	−0.0769	0.0188	1.375	1.0934	−0.196
90.00	2.375	1.5983	0.0954	0.3590	0.6049	2.4916	0.0788	−0.0940	−0.0769	−0.0142	1.771	1.0272	−0.055
95.00	1.295	1.7992	0.0520	0.1900	0.4399	2.6719	0.0591	−0.0907	−0.0721	−0.0346	2.380	1.0094	−0.019
97.50	0.790	1.8986	0.0317	0.0976	0.3162	2.8195	0.0503	−0.0840	−0.0657	−0.0500	3.250	0.9438	0.109
100.00	0.260	2.0000	0.0104	−0.0003	0.0706	3.0703	(0.0319)	−0.0606	−0.0467	−0.1150	41.850	1.1110	−0.233
0	0	−2.0167	0	−0.0167	−0.0173	6.2647	(0.1186)	0.0032	−0.0363	0.0060	63.900	−0.0064	1.000
1.25	−1.223	−1.9665	−0.0491	−0.0327	−0.2145	6.0658	0.1144	0.0625	−0.0478	0.0810	4.870	−1.0130	−0.025
2.50	−1.511	−1.9163	−0.0607	0.0811	−0.3022	5.9751	0.1001	0.1105	−0.0574	0.0382	3.560	−1.0808	−0.168
5.00	−1.942	−1.8158	−0.0780	0.1742	−0.4082	5.8615	0.0954	0.1010	−0.0419	0.0400	2.495	−1.0130	−0.026
7.50	−2.280	−1.7154	−0.0896	0.2642	−0.5190	5.7363	0.0862	0.0130	−0.0399	0.0072	2.076	−1.0392	−0.080
10.00	−2.550	−1.6150	−0.1012	0.3433	−0.5938	5.6465	0.0851	0.0097	−0.0420	0.0031	1.819	−1.0397	−0.081
15.00	−2.970	−1.4142	−0.1193	0.4964	−0.7095	5.4964	0.0839	0.0115	−0.0412	0.0014	1.412	−0.9671	0.065
20.00	−3.239	−1.2133	−0.1301	0.6277	−0.7965	5.3606	0.0815	0.0143	−0.0400	0.0018	1.370	−1.0550	−0.113
30.00	−3.527	−0.8117	−0.1417	0.8303	−0.9145	5.1276	0.0774	0.0200	−0.0399	0.0065	1.190	−1.0491	−0.101
40.00	−3.527	−0.4100	−0.1417	0.9330	−0.9789	4.9174	0.0722	0.0246	−0.0391	−0.0095	1.110	−1.0421	−0.086
50.00	−3.383	−0.0083	−0.1359	0.9954	−1.0005	4.7115	0.0678	0.0281	−0.0356	−0.0125	1.082	−1.0324	−0.066
60.00	−3.022	0.3933	−0.1214	0.9576	−0.9806	4.5152	0.0619	0.0325	−0.0314	−0.0156	1.101	−1.0246	−0.050
70.00	−2.519	0.7950	−0.1012	0.8394	−0.9178	4.3032	0.0551	0.0340	−0.0304	−0.0180	1.173	−1.0126	−0.025
80.00	−1.870	1.1967	−0.0751	0.6406	−0.8019	4.0711	0.0469	0.0360	−0.0252	−0.0280	1.454	−1.0848	−0.177
90.00	−1.079	1.5983	−0.0433	0.3608	−0.6020	3.7868	0.0360	0.0359	−0.0134	−0.0491	1.731	−0.9564	0.085
95.00	−0.648	1.7992	−0.0260	0.1905	−0.4375	3.5937	0.0297	0.0322	0.0032	−0.0638	2.345	−0.9300	0.135
97.50	−0.400	1.8986	−0.0161	0.0978	−0.3140	3.4603	0.0255	0.0255	0.0061	−0.0738	3.320	−0.9118	0.167
100.00	−0.160	2.0000	−0.0064	−0.0001	0	3.1416	0.0286	−0.0420	0.0367	−0.1715			1.000

Explanation of Table 8.1

1. Airfoil dimensions
2. Airfoil dimensions
3. From Eq. (8.12) and given leading-edge radius = 0.83 per cent chord. Hence

$$4a = 100 - \frac{0.83}{2}; \qquad a = 24.896 \text{ per cent chord}$$

$$x_1 = x - \left(2a + \frac{\rho}{2}\right) = x - 5.02075$$

4. $y_1 = y; a = 24.896$
5. From Eq. (8.5)
6. From Eq. (8.6)
7. From Col. 6
8. From Eq. (8.7)
9. Measured from plot of Cols. 7 and 8 (Fig. 8.3)
10. From Eq. (8.29) and Fig. (8.4)
11. Measured from Fig. 8.4
12. From Eq. (8.37). e^{ψ_0} from integrating the area under the ψ curve with a planimeter and using Eq. (8.25). ($\psi_0 = 0.095, e^{\psi_0} = 1.0995$)
13. From previous columns and Eq. (8.38), using $\alpha = +0.5°$
14. From Eq. 8.39

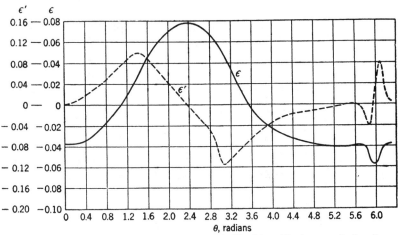

Fɪɢ. 8.4.   Plot of ϵ vs. θ needed to determine ϵ' in Ex. 8.1. ϵ' is shown as broken line.

From the ϵ curve we find that ϵ_T is 0.039, making the angle of zero lift $-2.23°$ (measured $-2.5°$), and the value of ϵ_N is $-.037$.   The ideal angle thus becomes essentially 0°.

Several pressure plots obtained by the Theodorsen method are presented in Fig. 8.5.   Using the discussion of Sect. 6.2, we find the airfoil acting at maximum efficiency in the range from $\alpha = 0.5$ to 1.6°, where extended laminar flow could develop on both surfaces.   Above $\alpha = 1.6°$ a high-velocity region exists near the upper-surface leading edge, and below the optimum range the lower surface has an adverse pressure gradient over most of the chord. Comparison of these pressure distributions with those of Figs. 6.8 and 6.17 indicates considerably lower drag in the optimum range for the modified 65,311 airfoil.

An additional point of interest is shown in Fig. 8.6, where pressure distributions for an NACA 0012-66 modified to have a slight camber are presented. Here it is illustrated that very special care must be taken

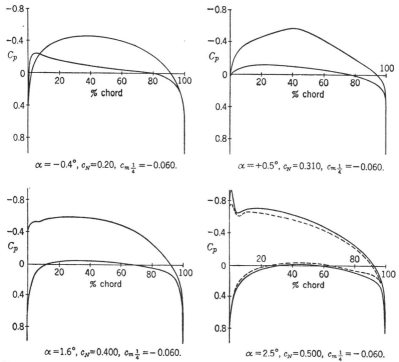

$\alpha = -0.4°$, $c_N = 0.20$, $c_{m\frac{1}{4}} = -0.060$.

$\alpha = +0.5°$, $c_N = 0.310$, $c_{m\frac{1}{4}} = -0.060$.

$\alpha = 1.6°$, $c_N = 0.400$, $c_{m\frac{1}{4}} = -0.060$.

$\alpha = 2.5°$, $c_N = 0.500$, $c_{m\frac{1}{4}} = -0.060$.

FIG. 8.5. Pressure plots by Theodorsen's method of the airfoil of Fig. 7.5. The broken lines in $\alpha = 2.5°$ are from experiment.

with airfoil thickness and camber distributions if good results are to be obtained. The airfoil of Fig. 8.6 is quite obviously an unsatisfactory one.

Some pressure distributions of an airfoil having smoother leading-edge pressure gradients are shown in Fig. 8.7.

8.2. Pinkerton's Addition to Theodorsen's Theory. At the time the Theodorsen theory was first published (1932) airfoils in common use were a long way from the low-drag designs now generally employed. As a consequence the theoretical lift curve slope used by Theodorsen was quite optimistic. Pressure distributions found by the Theodorsen method hence showed too great a lift for a specific angle of attack, or compared with practice, holding c_N constant, the shape of the theoretical and experimental distributions did not exactly agree. The difficulty was, of course, traceable to the boundary layer, and it was felt that the

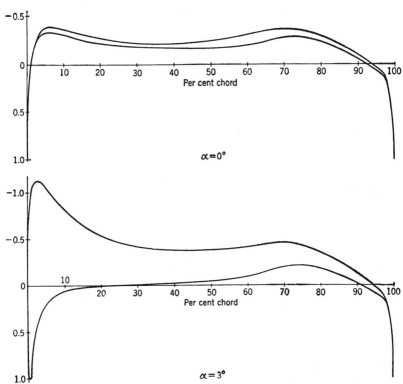

Fig. 8.6.   Pressure distribution over modified NACA 0012-66 airfoil.

Theodorsen theory would yield a better result if some method were found to use the coordinates of the airfoil plus boundary layer instead of those of the airfoil alone.

Pinkerton (Ref. 8.2) considered this gap between theory and practice and quite fortunately found a method that would account for the boundary layer without actually calculating its local effective thickness.   The first, and perhaps obvious, approach to the problem was to reduce the lift to agree with practice. This meant reducing the circulation for a given angle of attack. But reducing the circulation produced a violation of the Kutta-Joukowski law ("that the flow must leave the trailing edge

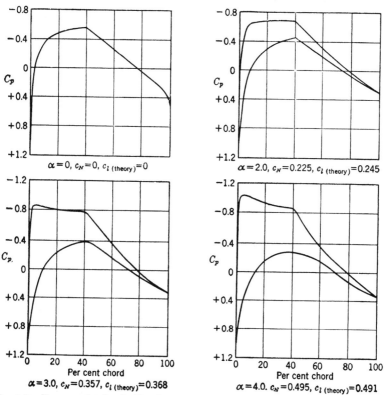

FIG. 8.7. Pressure plots by the Theodorsen method of an NACA 64, 3-018 airfoil. Experimental points closely check the above. (The C_N values are from integration of the curves above, showing reasonable error.)

smoothly"), and a large negative pressure peak appeared as the flow attempted to go round the sharp trailing edge.

The solution, as advanced by Pinkerton, employs the reduced circulation as obtained from an experimental lift curve, while excessive velocities at the trailing edge are avoided through an arbitrary modification to the ϵ shape-parameter curve that makes the velocity zero at $\theta = \pi$. In order to maintain a continuous ϵ curve, the arbitrary modification is spread out from the trailing edge sinusoidally. This puts the preponderance of the modification near the trailing edge, which fits in well with the known behavior of the boundary layer. The procedure is as follows:

First, follow through the Theodorsen method of evaluating θ, ψ, ψ_0, and ϵ. Next, from the lift curve as determined from force tests determine the actual lift developed at a selected α, and calculate the circulation from Eq. (4.26). Dividing both sides of Eq. (4.26) by $4\pi RV$, where $R = ae^{\psi_0}$, the radius of the z circle, we have

$$\frac{\Gamma}{4\pi RV} = \frac{cc_l}{8\pi R}$$

We next designate the altered ϵ curve as

$$\epsilon_\alpha = \epsilon + \frac{\Delta\epsilon_T}{2}(1 - \cos\theta) \qquad (8.41)$$

where $\Delta\epsilon_T$ is the increment of ϵ required to give zero velocity at $\theta = \pi$ and is a function of the angle of attack. The quantity $\Delta\epsilon_T$ is given by

$$\Delta\epsilon_T = \epsilon_{\alpha T} - \epsilon_T$$

where $\epsilon_{\alpha T}$ is determined by equating Eq. (8.37) to zero and substituting from Eq. (8.38). Hence

$$\sin(\pi + \alpha + \epsilon_{\alpha T}) - \frac{c}{8\pi R}c_l = 0$$

Solving for $\epsilon_{\alpha T}$, we have

$$\epsilon_{\alpha T} = \sin^{-1}\frac{c}{8\pi R}c_l - \alpha$$

where V = velocity of the undisturbed stream
$\quad\quad \alpha$ = angle of attack
$\quad\quad \Gamma$ = circulation
$\quad\quad \theta, \psi, \epsilon$ = airfoil shape parameters
$\quad\quad \psi_0$ = mean value of ψ
$\quad\quad R = ae^{\psi_0}$ = radius of conformal circle about which the flow is calculated

The parameters ϵ and ψ are conjugate functions of θ, and ψ is given by

$$\psi_n = \frac{1}{2\pi}\int_0^{2\pi}\epsilon\cot\frac{\theta - \theta_n}{2}d\theta + \psi_0$$

where the definite integral can be evaluated in the same manner as Eq. (8.28).

The influence of the changes in ψ on the value of F are found to be negligible. Defining

$$F = \frac{(1 + \epsilon')e^{\psi_0}}{\sqrt{(\sinh^2\psi + \sin^2\theta)[1 + (\psi')^2]}}$$

we write

$$F_\alpha = \left(1 + \frac{d\epsilon_\alpha}{d\theta}\right) F_1 \qquad (8.42)$$

where

$$F_1 = \frac{e^{\psi_0}}{\sqrt{(\sinh^2 \psi + \sin^2 \theta)[1 + (\psi')^2]}}$$

Differentiating Eq. (8.41), we have

$$\frac{d\epsilon_\alpha}{d\theta} = \frac{d\epsilon}{d\theta} + \frac{\Delta\epsilon_T}{2} \sin \theta$$

The equation for the velocity at any point is given by

$$v = VF_\alpha\left[\sin(\alpha + \theta + \epsilon_\alpha) + \frac{c}{8\pi R} c_l\right] \qquad (8.43)$$

and

$$C_p = 1 - \left(\frac{v}{V}\right)^2$$

The effect of using the Pinkerton modification in the calculation of the pressure distribution of the four-digit airfoils (NACA 0012, 2412, etc.)

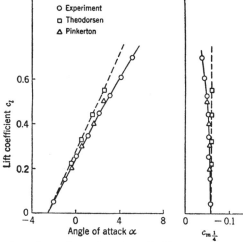

Fig. 8.8. Comparison of experimental results with Theodorsen and Pinkerton theories (example airfoil).

was found greatly to improve the agreement between theory and practice. And indeed, applying it to the airfoil of Fig. 7.5, for which the Theodorsen distributions were calculated earlier in the chapter, also gave better theory-practice agreement (see Fig. 8.8). The drawback to Pinkerton's method is the need of predetermining the lift curve slope. With airfoils of modern designs the agreement between the Theodorsen theory and practice is usually sufficient for engineering purposes.

Applying Eq. (8.3) to the modified ψ curve to obtain the coordinates of the altered airfoil sheds an interesting light on the necessity for the alteration or, correspondingly, the reason why many airfoils do not develop their theoretical lift curve slopes of slightly over 2π per radian. The altered airfoil, it develops, has a slightly turned-up trailing edge at a small positive angle of attack, and if the investigation be pursued further, it is seen that the amount of trailing-edge reflex increases with increasing angle of attack. Evidently the boundary layer thickens most on the upper surface, and the previously mentioned effective airfoil (the airfoil plus the effective boundary layer) is a shape that changes with angle of attack. This fits in well with logic, too, as after the minimum pressure point is reached, the rate of boundary-layer thickening is a function of the adverse pressure gradient against which it must advance. Since for lift to arise the upper surface of an airfoil must always have more negative pressures than the lower, it appears that the upper-surface boundary layer must always thicken faster than that of the lower surface, and accordingly the actual lift curve slope must always be less than the theoretical. Also, we should expect higher lift curve slopes from airfoils whose boundary layers are thin.

8.3. Airfoils with Specified Pressure Distributions. As we have seen, the Theodorsen theory will yield the pressure distribution around an arbitrary airfoil. If this pressure distribution is not one desired for a specific case, the cut-and-try method of Ref. 8.8 may be used to alter it (with some restrictions). But a direct solution to the inverse problem—that of finding an airfoil with a specific pressure distribution—may be made through the theory of Allen (Ref. 8.9). Allen's theory will not be discussed in detail since it follows principles already discussed in Chaps. 7 and 8 and can be employed without new knowledge.

The Allen theory takes advantage of the fact that the pressure distribution about an airfoil may be considered as due to a pressure distribution at the ideal angle plus an additional pressure distribution. Hence a perfectly arbitrary pressure distribution could be due to any of a large number of airfoils, each at a different angle of attack. However, as Allen points out, when a pressure distribution is selected for a proposed airfoil, the chances are overwhelming that it will be selected at its ideal angle, since there is little reason to want a particular airfoil at any other angle. Thus the problem is reduced to one with a unique solution whose basic steps are as follows:

1. Search the plots of velocity distribution of known airfoils (as, for instance, in Ref. 7.6), and find two airfoils whose velocity distributions

correspond as nearly as possible to the upper- and lower-surface velocity distributions desired. Average the nose radii of the two airfoils, and select as a "reference" airfoil a Joukowski airfoil which has the same nose radius as the above average.

2. Average the desired upper- and lower-surface velocity distributions to obtain the no-camber velocity distribution desired. Note the velocity difference between the reference airfoil of step 1 above and the mean of upper- and lower-surface values. Plot this velocity difference against θ (defined as usual), and adjust the curve so obtained until

$$\int_0^\pi \frac{\Delta v}{V}\, d\theta = 0$$

and

$$\int_0^\pi \frac{\Delta v}{V} \cos\theta\, d\theta = 0$$

which assures a real, closed airfoil.

3. Using the adjusted $\Delta v/V$ curve and the relation

$$\frac{d\,\Delta y}{dx} = \frac{1}{2\pi}\int_0^{2\pi} \frac{\Delta v}{V} \cot \frac{\theta - \theta_0}{2}\, d\theta$$

perform a strip integration similar to the integration in Appendix 2 to obtain $d\,\Delta y/dx$, which is then plotted against the chord station x. This curve is planimetered to yield Δy. The ordinates of the symmetrical no-camber shape at zero angle of attack are obtained from

$$y = y_r + \Delta y$$

where y_r is the ordinate of the previously selected reference airfoil.

4. The shape of the camber line is determined through the use of Eq. (7.18). Thus using the difference between the average velocity distribution and the upper- and lower-surface velocity distributions, obtain the value of P_b from

$$P_b = 2\left(\frac{V_u}{V} - \frac{V_l}{V}\right)$$

Plot $P_b/4$ against θ, and using the relation

$$\frac{dy_c}{dx} = \frac{1}{2\pi}\int_0^{2\pi} \frac{P_b}{4} \cot \frac{\theta - \theta_0}{2}\, d\theta$$

and the strip integration method find dy_c/dx and plot it against chord station x. Planimetering then yields the values of y_c, which must be then corrected to customary airfoil coordinates.

5. Add the ordinates of the camber line to those of the symmetrical no-camber shape to obtain the final airfoil.

As noted above, the Allen method requires more time than a Theodorsen solution since two curves must be plotted and their slopes measured. This extra work might be inferred since two solutions—that for the symmetrical shape and that for the camber line—are required.

8.4. Estimation of Airfoil Drag. In the preceding pages we have discussed methods of determining the theoretical angle of zero lift, slope of the lift curve, and moment of an airfoil and found that either the thin-airfoil theory or Theodorsen's theory are strong approaches—suitable in many cases for engineering requirements. But in no case has a drag arisen. This, of course, is at wide variance with practice.

The study and application of viscous-fluid theory is beyond the scope of this textbook and hence will not be presented. However, a simple, very approximate method useful for first-order work is given below. This method consists in assuming that the drag of an airfoil *in its low-drag range* is composed only of skin friction and further that that drag is the same as the well-known values for constant-pressure regions:

$$c_{dl} = \frac{1.328}{\sqrt{RN}} \qquad \text{for a laminar boundary layer} \qquad (8.44)$$

$$c_{dt} = \frac{0.455}{(\log_{10} RN)^{2.58}} \qquad \text{for a turbulent boundary layer} \qquad (8.45)$$

Under these limited conditions the assumption of zero form drag does not seem amiss. The procedure to follow for calculating the drag in the optimum range is outlined below:

1. At a particular angle of attack examine the pressure-distribution diagram, and assume that laminar flow will exist up to the minimum pressure points on upper and lower surfaces and that turbulent flow will exist thenceforth to the trailing edge.

2. At the desired airfoil Reynolds number R_1 calculate the drag coefficients for the laminar and turbulent parts, using Eqs. (8.44) and (8.45). The Reynolds number of the laminar part (x per cent chord) will be xR_1, while the Reynolds number of the turbulent part will be $R_1 - (xR_1/2)$ to allow for the transition region.

3. The airfoil drag coefficient will finally be

$$c_{do} = \left(\frac{x}{100} c_{dl} + \frac{100 - x}{100} c_{dt} \right)_{\text{upper surface}} + \left(\frac{x}{100} c_{dl} + \frac{100 - x}{100} c_{dt} \right)_{\text{bottom surface}}$$

An example of the method is presented below, while a comparison with experiment is shown in Fig. 8.9.

Example 8.2. Estimate the drag of the airfoil shown in Fig. 7.5 at an angle of attack of 0.5° at RN = 2, 200,000. The chord is 2 ft.

1. From Fig. 8.5 it is estimated that laminar flow will exist to the 40 per cent chord point on the upper surface and to the 20 per cent point on the lower. Thus the Reynolds numbers for the laminar flow are 880,000 and 440,000 and for the turbulent flow 1,760,000 and 1,980,000

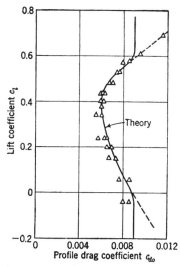

FIG. 8.9. Comparison of approximate drag estimation and experiment (example airfoil).

2. The laminar drag coefficients are

$$c_{dlu} = \frac{1{,}328}{\sqrt{880{,}000}} = 0.00141$$
$$c_{dlb} = 0.00200$$

3. The turbulent drag coefficients are

$$c_{dtu} = \frac{0.455}{(\log_{10} 1{,}760{,}000)^{2.58}} = 0.00406$$
$$c_{dtb} = 0.00399$$

4. $c_{d0} = 0.4(0.00141) + 0.2(0.00200) + 0.6(0.00406) + 0.8(0.00399) = 0.0066$

The measured drag coefficient was 0.0062. The significance of having experiment less that theory merely indicates that (1) the laminar flow actually extends slightly beyond the minimum pressure peaks and (2) the formulas for constant pressure regions may not be exactly applicable to regions of varying pressure.

REFERENCES

8.1. Theodore Theodorsen, Theory of Wing Sections of Arbitrary Shape, *TR* 411, 1932.

8.2. Robert M. Pinkerton, Calculated and Measured Pressure Distributions over the Midspan Section of an NACA 4412 Airfoil, *TR* 563, 1936.

8.3. E. B. Wilson, "Advanced Calculus," Ginn & Company, Boston.

8.4. I. Naiman, Numerical Evaluation of the ϵ Integral Occurring in the Theodorsen Arbitrary Airfoil Potential Theory, *WRL* 136, 1944.

8.5. I. Naiman, Numerical Evaluation by Harmonic Analysis of the ϵ Function of the Theodorsen Arbitrary Airfoil Potential Theory, *WRL* 153, 1945.

8.6. Theodore Theodorsen and I. E. Garrick, General Potential Theory of Arbitrary Wing Sections, *TR* 452, 1933.

8.7. Alan Pope, On Airfoil Theory and Experiment, *JAS*, July, 1938.

8.8. Theodore Theodorsen, Airfoil Contour Modifications Based on ϵ-curve Method of Calculating Pressure Distribution, *WRL* 135, 1944.

8.9. H. Julian Allen, General Theory of Airfoil Sections Having Arbitrary Shape of Pressure Distribution, *TR* 833, 1945.

8.10. Ira H. Abbott and Albert E. von Doenhoff, "Theory of Wing Sections," McGraw-Hill Book Company, Inc., New York, 1949.

CHAPTER 9

THE FINITE WING

9.1. The Finite Wing. Up to this point we have considered only two-dimensional flow and have carried that development up to its current status as a coherent phase of aerodynamics. Now we shall turn our attention to the finite wing and see what effect freeing the tips has on its performance.

The basic mechanism of flight was understood and practiced (with gliders) a long time ago. Many theorists became interested in its problems, and a decade before the first powered flight Lanchester (born 1868) in England postulated the type of flow that would be experienced by a finite wing. In a paper in 1894 which later led to the publication of his book "Aerodynamics"* in 1907 (Ref. 9.1) he stated that the high pressure beneath the wings would spill out around the tips into the low-pressure region above the wings, forming vortices which would stream out behind the wing. These vortices would (he continued) roll up into two main trailing vortices of opposite sign, located one behind each tip, and would be deflected downward. Further, their effect upon the flow at the wing would tip the resultant force vector back, causing a component of the lift to become *induced drag*. It is evident that Lanchester clearly understood this phenomenon when it is noted that in 1897 he secured a patent covering the use of end plates at the wing tips to minimize the *spillage* there—six years before the Wright brothers' flight!

Although Lanchester's work was mathematical in scope, his presentation was not, and it remained to Prandtl (born 1875) to extend the work of Lanchester into the *Prandtl lifting-line theory* (Ref. 9.2) presented in 1911 and developed in the following pages.

9.2. The Trailing Vortices. As Lanchester pointed out, the important difference in the flow pattern about two- and three-dimensional wings is traceable to the difference in spanwise lift distribution, which is in turn traceable to the disposition of the circulation. The two-dimensional wing has constant circulation along the *span*. When we represent the airfoil by a vortex, its effect ahead of the wing tending to increase the angle of attack is exactly balanced by the effect behind the wing tending to decrease it.

---

* Especially recommended to the student for its historical value.

181

On the other hand, the three-dimensional wing has a diminution of circulation toward the wing tip until, when the tip is reached, the circulation is zero.   At each change of circulation a vortex is shed, and the shed vortices, being of similar sign, roll up into a single trailing vortex somewhat inside each wing tip.   The net result is that the vortex pattern behind the wing is no longer similar to that ahead of the wing.   The

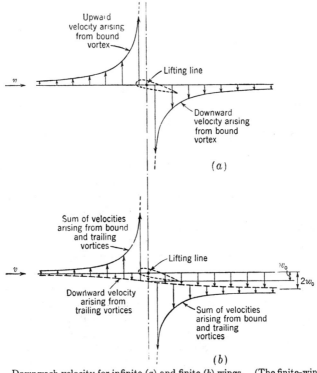

FIG. 9.1.   Downwash velocity for infinite (a) and finite (b) wings.   (The finite-wing value of $2w_0$ is for a first-approximation elliptic wing.)   (*Reproduced by permission from "Principles of Aerodynamics" by James H. Dwinnell, published by McGraw-Hill Book Company, Inc.,* 1949.)

vertical induced velocities ahead and behind the wing no longer cancel, as the air behind the wing is affected by the downthrust on the wing circulation *and* the trailing vortices, while the air ahead of the wing is affected almost entirely by the upward component of the wing circulation (see Fig. 9.1).   (The effect of the trailing vortices is very small ahead of the wing.)   Finally, the whole field behind the wing has a downward inclination whose vertical component is called the *downwash,* and at the

THE FINITE WING 183

wing enough inclination results materially to tip the resultant force backward and create a drag.

A great contribution to the theory for a complete wing was Prandtl's statement that each section of the wing acts as though it is an isolated two-dimensional section at an angle of attack α_0. This condition, known as *Prandtl's hypothesis*, is in general responsible for the existence of a simple theory for a complete wing of finite span. There are, however, certain conditions when the spanwise flow becomes so pronounced that it is not applicable, as, for instance, when severe lateral pressure gradients arise with a swept-back wing. Since these conditions become of interest in determining the stalling characteristics of wings, they offer an irritating obstacle to practical design at the current

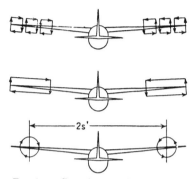

FIG. 9.2. Growth of shed vortices into tip vortex, showing cancellation of components and final reduced span.

time. They are due, of course, to the fact that curved flow ensues, owing to the above reasons, and the actual effective airfoil sections are undefined.

Another great contribution to the mathematical representation of this field was Prandtl's statement that the trailing vortices could be con-

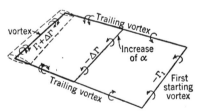

FIG. 9.3. Representation of starting vortex and a change of lift.

sidered as part of one vortex which extends across the wing span and is closed a long way behind the wing by a *starting vortex* (Fig. 9.3 and Sect. 4.2). This complete vortex representation will enable us to calculate the induced velocity at any point near a wing as soon as we develop relations to parallel in three dimensions the simple relation of Eq. (4.3).

But first a few words about this *three-dimensional vortex*, which now has a finite length to the axis of rotation. Our previous vortex was flat (two-dimensional) and its core a disk. The concept of three dimensions requires a vortex of finite length and, of course, a finite-length core as well.

Helmholtz (1821–1894) in 1858 formulated vortex laws that will assist us in studying the three-dimensional vortex as well as the vortex pattern of a finite wing. Briefly these were:

1. Vortex filaments either form closed curves or extend to the fluid boundaries.

2. The circulation remains constant throughout the length of a vortex.

3. The particles of fluid composing a vortex remain with that vortex indefinitely.

Sir William Thomson (Lord Kelvin, 1824–1907) is usually credited with the additional law, "Circulation remains constant with time."

Let us see how well actual conditions meet these laws.

Flow pictures made (Ref. 9.4) at the instant air begins to move over a wing at some lifting angle of attack show that a vortex, parallel to the wing span, leaves the wing trailing edge and proceeds downstream. This vortex is called the *starting* vortex and is of strength equal to that of the *lifting*, or *bound*, vortex. The sum of the two is thus zero, satisfying the Thomson law.

The starting and bound vortices may be very simply demonstrated with a razor blade and a basin of water. The razor blade, held half immersed so that it represents a wing at some lifting angle, is moved briskly through the water. A starting vortex immediately leaves the blade. As long as the motion is constant, no additional vortices appear, but any increase of angle produces a second starting vortex. When the motion is stopped, the bound vortex comes off the "wing." The starting vortex, by connecting the ends of the two trailing vortices of the wing vortex system, enables that system to meet the first Helmholtz law. The system meets the second law by actually having constant circulation in each vortex section. The third law is not quite satisfied, since the system is continually picking up air and thrusting it into the trailing vortex system. Indeed, in view of the molecular action taking place in any gas there is doubtless a considerable exchange of molecules between those in the core and in free air.

After the starting vortex has moved downstream to a point where its influence near the wing is negligible, there remains only the bound vortex with two trailing vortices in the region. The shape of this combination resembles a horseshoe, and ordinarily the system is so called, instead of identifying its components.

Since the lift vector of a circular cylinder goes through its axis, the horseshoe vortex system is most accurate if the bound vortex is assumed to run through the wing center-of-pressure line.

We see directly that our simple horseshoe vortex field, having its vortex strength determined from $L = \rho V \Gamma b$, fits in well with the actual conditions. Silverstein, Katzoff, and Bullivant (Ref. 11.1) examined this phenomenon by comparing the flow field about a Clark Y as determined by the method of Chap. 8 and the vortex laws of Chap. 4. It was discovered that at a distance of one chord from the wing trailing edge the downwash angles were matched within 0.3° by a simple point vortex field.

While we have just discussed the case of constant circulation across the span, it may be seen by reference to Fig. 9.4 that any symmetrical (or, indeed, unsymmetrical) spanwise distribution of lift may be simulated by a number of superimposed closed vortex systems. In actuality, these small vortices, shed at every change of circulation on the span, roll up into the tip vortices and move them slightly toward the plane of

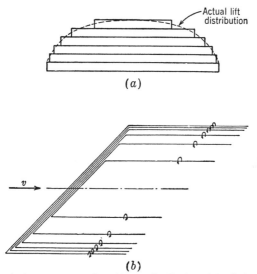

(a)

(b)

FIG. 9.4. Theoretical vortex pattern for arbitrary distribution of circulation along the span, roll-up of shed vortices neglected. (*Reproduced by permission from "Principles of Aerodynamics" by James H. Dwinnell, published by McGraw-Hill Book Company, Inc., 1949.*)

symmetry. In addition the whole field aft of the wing is displaced vertically downward by the bound vortex.

The manner in which the circulation varies along the span is of great interest to the aerodynamicist since it will be shown that one particular variation (elliptic) results in *minimum* induced drag, others producing from 2 to 15 per cent more. All variations of the circulation are amenable to mathematical treatment, but for simplicity we shall consider four: elliptic (produced by an untwisted wing with elliptic planform; uniform (not actually produced by any planform, but of such simple mathematical scope that it helps in understanding the other cases); rectangular-wing type; and tapered-wing type. It is emphasized that rectangular wings do *not* have uniform loading, for reasons that will be brought out later.

9.3. The Biot-Savart Law. Before we can actually calculate the effect of the trailing vortices, it is necessary to develop vortex laws for three-dimensional flow. Added value may be attached to this particular development as it embraces nearly every aerodynamic and vector principle studied so far.

First let us recall some of the rules and definitions of the velocity potential. ϕ has been defined by the relation

$$\phi = \int_0^P q \cos \theta \, ds = \int_0^P \mathbf{q} \cdot d\mathbf{s} \tag{9.1}$$

where the symbols are shown in Fig. 4.11.

The velocity potential of one point A relative to another point A' is

$$\phi_{A'} - \phi_A = \int_0^{A'} \mathbf{q} \cdot d\mathbf{s} - \int_0^A \mathbf{q} \cdot d\mathbf{s} \tag{9.2}$$

$$\phi_{A'} - \phi_A = \int_A^{A'} \mathbf{q} \cdot d\mathbf{s} \tag{9.3}$$

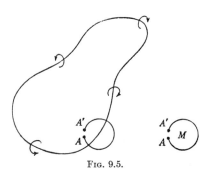

FIG. 9.5.

It was demonstrated in Sect. 4.6 that ϕ exists only for irrotational motion and is then independent of the path of integration—in this case AA' in Fig. 9.5. It is further noted in Chap. 4 that the value of the circulation bears a close resemblance to Eq. (9.3):

$$\Gamma = \oint \mathbf{q} \cdot d\mathbf{s} \tag{9.4}$$

The only difference is that for circulation the path of integration is closed. Under these conditions, if the line integral about the closed curve is *not* zero, its value *is* the circulation, and the motion is not *completely* irrotational unless a solid body is included.

Now consider an arbitrary surface which has a vortex for a boundary. The line integral along any closed path M that does not cut the surface is zero. But when the surface is cut, we get a sudden rise in potential due to the vortex. It is advantageous to assume a thickness to the surface to avoid an infinite rate of rise (see Fig. 9.6).

Now a given potential may be produced by sources and sinks or by vortices. At this point it is more convenient to consider the flow field produced by the closed vortex line of Fig. 9.5 as being produced by a double sheet of sources and sinks as shown in Fig. 9.7. For instance, if Q is the source quantity, then the velocity at point P a distance r

away will be

$$q = \frac{Q}{4\pi r^2} \tag{9.5}$$

[Compare Eq. (9.5) with the development of Eq. (3.13).]

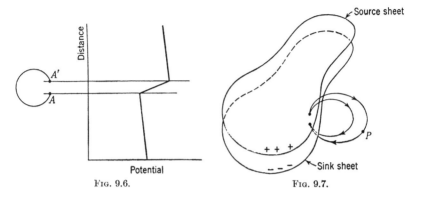

FIG. 9.6.

FIG. 9.7.

The potential at point a due to a source at the origin is

$$\phi_a = \int_0^a \mathbf{q} \cdot d\mathbf{s} \tag{9.6}$$

which in this case is

$$\phi_a = \int_0^a \mathbf{q}_r \cdot d\mathbf{r} \tag{9.7}$$

But q_r is straight (the flow from a source is radial), and the angle between q_r and dr is zero (Fig. 9.8), so that

$$\phi_a = \int_0^a q_r \, dr$$
$$= \int_0^a \frac{Q}{4\pi r^2} \, dr = -\frac{Q}{4\pi a} \tag{9.8}$$

or for a sink

$$\phi_a = \frac{Q}{4\pi a}$$

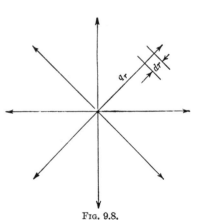

FIG. 9.8.

If we call the source intensity Q_1, the potential at P due to elemental area dA of the doublet is

$$d\phi = -\frac{Q_1 \, dA}{4\pi a_o} + \frac{Q_1 \, dA}{4\pi a_i} \tag{9.9}$$

where a_o and a_i are the distances from P to the proper source and sink elements.

We see, however (Fig. 9.9), that

$$a_o - h_1 \cos \alpha = a_i \tag{9.10}$$

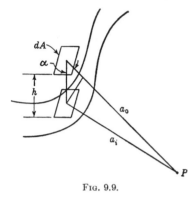

FIG. 9.9.

where h_1 is the distance between source and sink sheets, and hence

$$\frac{1}{a_o} = (a_i + h_1 \cos \alpha)^{-1}$$

$$= \frac{1}{a_i} - \frac{h_1 \cos \alpha}{a_i^2} + \frac{h_1^2 \cos^2 \alpha}{a_i^3} \tag{9.11}$$

Excluding the last term of Eq. (9.11) as small, we have

$$\frac{1}{a_i} - \frac{1}{a_o} = \frac{h_1 \cos \alpha}{a^2} \tag{9.12}$$

We may drop the subscript on the a^2 term as with h_1 small and $a \gg h_1$, $a_i \cong a_o$ and in the term where a^2 appears the difference between a_i and a_o is inconsequential.

Equation (9.9) is then

$$d\phi = \frac{Q_1 \, dA}{4\pi} \cdot \frac{h_1 \cos \alpha}{a^2} \tag{9.13}$$

Now consider the flow between upper and lower sheets. Except for a very small (but important) amount, the entire flow is from source to sink between the sheets. For a square-foot area the quantity is Q_1 cu ft per sec, and hence the velocity developed from upper to lower sheet is numerically equal to Q_1 ft per sec. As this takes place along a distance h_1, we see the rise in potential = velocity × distance = $Q_1 h_1$.

As we have defined the flow as producing a rise in potential equal to the vortex strength (or Γ), then $Q_1 h_1 = \Gamma$ and Eq. (9.13) becomes

$$d\phi = \frac{\Gamma}{4\pi} \frac{\cos \alpha \, dA}{a^2} \tag{9.14}$$

Equation (9.14) relates the velocity potential with the circulation but is not yet in a usable form. Indeed, from a practical standpoint, we shall want to change it to another general form.

Now, return to the vortex-sheet concept, and put a small unit spherical surface about the point P (Fig. 9.10a). From Fig. 9.10b the projection of dA perpendicular to a is essentially $dA \cos \alpha$, and, assuming $dA \cos \alpha$ to be square, the rays from dA to P will define a small "square" (or solid angle ω) on the surface of the sphere. On the unit sphere the projection will be inversely proportional to the square of the distance, or

$$d\omega = \frac{dA \cos \alpha}{a^2} \tag{9.15}$$

From Eq. (9.14)

$$d\phi = \frac{\Gamma}{4\pi} \frac{dA \cos \alpha}{a^2} = \frac{\Gamma}{4\pi} d\omega \tag{9.16}$$

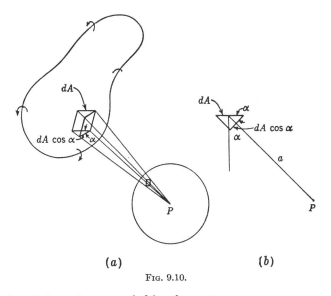

(a) (b)

Fig. 9.10.

or for the whole surface bounded by the vortex

$$\phi = \frac{\Gamma}{4\pi} \omega \tag{9.17}$$

Now the velocity at point P is

$$\begin{aligned}
\mathbf{q} &= iu + jv + kw \\
&= i\frac{\partial \phi}{\partial x} + j\frac{\partial \phi}{\partial y} + k\frac{\partial \phi}{\partial z} \\
&= \nabla \frac{\Gamma}{4\pi} \omega = \frac{\Gamma}{4\pi} \nabla \omega \tag{9.18}
\end{aligned}$$

We seek $\mathbf{q}$ due to the vortex of strength Γ, and our development so far tells us that $\mathbf{q}$ is a function of the solid angle ω. It is apparent that ω would change if the point P were moved from the surface or, equally, if the vortex surface were moved from the point. The latter approach, it develops, is easiest to analyze. We must move the sheet along a, holding it parallel to itself, as any angular motion would correspond to rotating the point P about the sheet, and we wish to examine the effect of only *moving P*.

From inspection, increasing the distance a decreases the potential at P as it decreases ω. Thus it is that while we consider the effect of moving the vortex sheet we are also considering the effect on the potential at P.

From Fig. 9.11, the only movement of the vortex sheet that affects ω will be that which affects the projected area $dA_1 = dA \cos \alpha$. Moving

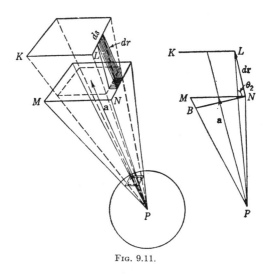

FIG. 9.11.

the sheet parallel to itself, we see the net result is a decrease of ω by a perimetric strip, and we may evaluate this strip as follows: The displacement along a of the vortex sheet parallel to itself creates a surface of height dr and perimeter $\oint ds$. A section of this surface is $ds\, dr$, and the projection of $ds\, dr$ on the perpendicular to a is $ds\, dr \cot \theta_2$, where θ_2 is the angle between the radius vector PL and a perpendicular to a, BN.

Now dr is perpendicular to ds, and the vector that represents the area $ds\, dr$ may hence be written $\epsilon\, ds\, dr$ or $d\mathbf{s} \times d\mathbf{r}$. The vector that represents the area dA_1 will be perpendicular to dA_1 and hence lie along a and have a unit length $a_1 = \mathbf{a}/a$. In addition the angle θ_2 will be nearly $90°$, and we may write $\cot \theta_2 = \cos \theta_2$. Finally the change in projected area as the sheet is moved becomes

$$\Delta(dA_1) = d\mathbf{r} \times d\mathbf{s} \cdot \frac{\mathbf{a}}{a} \qquad (9.19)$$

and the change in the solid angle ω is then

$$d\omega = \frac{dA_1}{a^2} = \oint \frac{d\mathbf{r} \times d\mathbf{s} \cdot \mathbf{a}}{a^3} \tag{9.20}$$

which by Eq. (1.25) becomes

$$d\omega = \oint \frac{d\mathbf{r} \cdot d\mathbf{s} \times \mathbf{a}}{a^3} \tag{9.21}$$

As we have moved the vortex sheet parallel to itself only, the direction and length of $d\mathbf{r}$ is constant. Hence we may remove $d\mathbf{r}$ from the integral [Eq. (9.21)] as follows:

$$d\omega = d\mathbf{r} \cdot \oint \frac{d\mathbf{s} \times \mathbf{a}}{a^3} \tag{9.22}$$

To digress a moment,

$$d\mathbf{r} \cdot \nabla\omega = \frac{\partial \omega}{\partial x} dx + \frac{\partial \omega}{\partial y} dy + \frac{\partial \omega}{\partial z} dz \tag{9.23}$$

From the definition of a derivative we then have

$$d\mathbf{r} \cdot \nabla\omega = d\omega$$

So from Eq. (9.23)

$$d\mathbf{r} \cdot \nabla\omega = d\mathbf{r} \cdot \oint \frac{d\mathbf{s} \times \mathbf{a}}{a^3} \tag{9.24}$$

and

$$\nabla\omega = \oint \frac{d\mathbf{s} \times \mathbf{a}}{a^3}$$

Substituting in Eq. (9.18),

$$\mathbf{q} = \frac{\Gamma}{4\pi} \oint \frac{d\mathbf{s} \times \mathbf{a}}{a^3}$$

and substituting for $d\mathbf{s} \times \mathbf{a}$ its equal, $\epsilon a\, ds \sin\theta$, where θ is the angle between a and the vortex segment ds, we have (see Fig. 9.12b)

$$|\mathbf{q}| = \frac{\Gamma}{4\pi} \int_A^B \frac{ds \sin\theta}{a^2} \tag{9.25}$$

Now, having the expression for the effect of a segment of vortex, we may proceed to the application problems. First, however, it is convenient to express Eq. (9.25) in trigonometric terms.

Taking out a section $a\, d\theta$ (Fig. 9.12a), we see that $dx = a\, d\theta$;

$$\sin\theta = dx/ds$$

so that

$$ds = \frac{a\, d\theta}{\sin\theta}$$

Also from Fig. 9.12b $\sin \theta = h/a$, so that

$$ds = \frac{h \, d\theta}{\sin^2 \theta} \tag{9.26}$$

Changing limits, when $s = A$, $\theta = \theta_1$ and when $s = B$, $\theta = \theta_2$, and, substituting, Eq. (9.25) becomes

$$|\mathbf{q}| = \frac{\Gamma}{4\pi h} (\cos \theta_2 - \cos \theta_1)$$

Considering the two interior angles as easier to use, we have

$$|\mathbf{q}| = \frac{\Gamma}{4\pi h} (\cos \theta_1 + \cos \theta_3) \tag{9.27}$$

Equation (9.27) is a common form of the Biot-Savart law. It demonstrates first of all that, as the vortex segment is approached, the induced

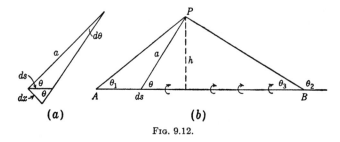

Fig. 9.12.

velocity it produces rapidly increases. Now, according to the Helmholtz laws the vortex cannot abruptly end in the fluid, but when it turns a sharp corner the usual approach is to consider each straight segment separately, just as though it did terminate at the corner. Thus for a vortex that "starts" at P and extends to infinity out beyond B, $\theta_1 = \pi/2$, and $\theta_3 = 0$, and

$$q = \frac{\Gamma}{4\pi h} \tag{9.28}$$

Or if the vortex extends to infinity in both directions, $\theta_1 = \theta_3 = 0$ and

$$q = \frac{\Gamma}{2\pi h} \tag{9.29}$$

Of interest is the fact that the induced velocity exists beyond the "end" of a vortex, although its value is small (Fig. 9.13).

The replacing of a wing by a simple lifting-line vortex and the proper use of the Biot-Savart law are illustrated in subsequent chapters.

9.4. Downwash at the Wing for Any Spanwise Distribution of Circulation.

Undoubtedly the most important effect of the trailing vortices

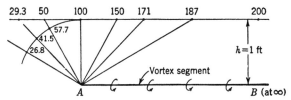

29.3 50   100   150 171   187   200

57.7
41.5
26.8

$h = 1$ ft

Vortex segment

A   B (at ∞)

Fig. 9.13. Values of induced velocity near the "end" of a vortex segment, $\Gamma = 400\pi$ sq ft per sec.

is the increase of drag they produce through altering the flow angle at the wing. Secondarily, the *manner* in which they alter the flow at the wing is important, too, as usually the induced velocity is not constant across the span, and the local angle of attack varies over a large range.

Let us call the downwash velocity w and the wing span b. Orient the wing so that the x axis coincides with the forward airplane longitudinal axis, the y axis coincides with the bound, or lifting, vortex, and the z axis is downward. (Downwash is positive in the downward direction.)

Since the circulation decreases toward the tips (it must be zero at the

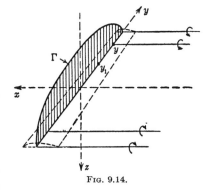

Fig. 9.14.

tip), between the points y and $y + dy$ on the span the circulation changes $-\dfrac{\partial \Gamma}{\partial y} dy$. Hence the induced velocity at the point y_1 for a differential vortex at y is, from Eq. (9.28) and Fig. 9.14,

$$dw(y_1)^* = \frac{-\dfrac{\partial \Gamma}{\partial y} dy}{4\pi(y - y_1)} \tag{9.30}$$

and the total downwash at y_1 will be

$$w(y_1) = \frac{1}{4\pi} \int_{-b/2}^{b/2} \frac{\dfrac{\partial \Gamma}{\partial y} dy}{(y_1 - y)} \tag{9.31}$$

* $dw(y_1)$ should be read "dw at y_1."

Equation (9.31) hence defines the amount of downwash that will occur at point y_1 for any arbitrary distribution of circulation described by $\dfrac{\partial \Gamma}{\partial y}$. It is known as the *Prandtl downwash relation*.

Using the local downwash w, we may then find the local angle of attack and local lift and finally the approximate span load distribution. Repeating the process, we could in theory find the actual span loading by eliminating the effect of each new downwash distribution on the original assumed span loading. However, the problem is extremely complex, and we shall usually confine ourselves to considering a uniform or an elliptic span loading in this and the following chapter. (In Chap. 12 two methods for getting the span loading for arbitrary planforms are discussed.)

9.5. Downwash at the Wing for Elliptic Distribution of Circulation. Before proceeding further it is in order to discuss the terms *span loading*, *spanwise lift distribution*, and *spanwise lift coefficient distribution*. Since

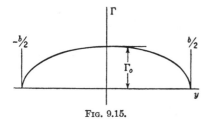

FIG. 9.15.

[Eq. (4.4)] the lift is a function only of the density, velocity, and circulation, for a given set of conditions an elliptic distribution of circulation would also be an elliptic distribution of lift. However, the relation between the section lift and the section lift coefficient involves the local chord, and hence elliptic lift or distribution of circulation may or may not mean elliptic distribution of c_l. In fact, the only case where all three distributions are elliptic is that of the properly twisted rectangular wing.

Proceeding now to the case of elliptic loading ("loading" as we have seen may be either lift or circulation), we assume a wing that has a circulation Γ_0 at mid-span which decreases elliptically until it is zero at the wing tips (see Fig. 9.15). We write

$$\frac{y^2}{(b/2)^2} + \frac{\Gamma^2}{\Gamma_0{}^2} = 1 \tag{9.32}$$

or

$$\Gamma = \Gamma_0 \sqrt{1 - \left(\frac{2y}{b}\right)^2} \tag{9.33}$$

Then

$$\frac{\partial \Gamma}{\partial y} = \frac{-\Gamma_0 y}{(b/2)\sqrt{(b/2)^2 - y^2}} \tag{9.34}$$

Substituting into Eq. (9.31), we get that the downwash at point y_1 on the span is

$$w(y_1) = \frac{1}{4\pi} \int_{-b/2}^{b/2} \frac{-\Gamma_0 y \, dy}{(y_1 - y)(b/2) \sqrt{(b/2)^2 - y^2}} \tag{9.35}$$

The integrations of Eq. (9.35) is complicated by the fact that the value of the integrand is infinite when $y = y_1$, $b/2$ or $-b/2$. The complete integration is given in Appendix 3, and we may hence state

$$w(y_1) = \frac{\Gamma_0}{2b}$$

But y_1 is any point on the span, and hence this relation applies to every point on the span.

$$w(y) = \frac{\Gamma_0}{2b} \tag{9.36}$$

The interesting conclusion is that, if the distribution of circulation is elliptic across the span, the downwash at the wing due to the trailing vortices is constant across the span. While we might rightfully conclude that the local angle of attack is also constant for an untwisted elliptic wing, the conclusion cannot be extended to include a simultaneous stall across the span, as practical considerations of viscosity dictate an earlier stall for the shorter chord, lower Reynolds number wing tips.

9.6. Downwash at the Wing for Uniform Distribution of Circulation. The case of uniform distribution of circulation is developed in Chap. 11, but we may state the results here to compare with Sect. 9.4. It is shown in Chap. 11 that the downwash distribution at a wing with uniform loading is

$$w = -\frac{\Gamma}{2\pi} \frac{s(y^2 - s^2)}{(y^2 + s^2)^2 - 4s^2 y^2} \tag{9.37}$$

(The symbols are defined in Chap. 11.)

Substituting for example $\Gamma = 6.28$ and $s = 1.0$, we get the values of w shown in Fig. 9.16. Here it is seen that the angle of attack is locally reduced by uniform loading all along the span, but unlike the case of elliptic loading the reduction is not uniform, being greatest at the tips. This tends toward a wing root stall appearing earliest. We may also get nearly uniform loading by twisting a tapered wing, but in practice the twist is kept small to avoid excessive deviations from the elliptic-type loading, for reasons to be advanced later.

As a matter of interest the effect of the trailing vortices on the uniformly loaded wing quite obviously tends toward destroying its uniformity of lift, and hence even the rectangular wing tends toward an

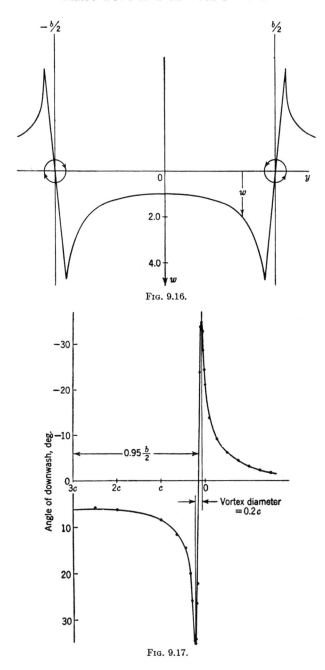

Fig. 9.16.

Fig. 9.17.

elliptic distribution. Experience tells us it gets about halfway there, but the alteration from the basic uniformity is not enough to alter the expected root stall.

An interesting check of this theory and of Fig. 9.16 was made by Piercy and is reported in Ref. 9.3. In these experiments a wing of 12 in. span and 3 in. chord was set at 8° in a wind tunnel with a 4-ft-square test section. Using a small yawhead, the angle of flow was measured $0.9b$ behind the wing in a plane near that of the wing. The results, shown in Fig. 9.17, demonstrate clearly the existence of a vortex similar to that supposed by the theory. At a distance of $0.5c$ behind the trailing edge a trailing vortex was clearly discerned. Its core had a diameter of $0.2c$, and its center was $0.95s$ from the plane of symmetry. Inside the core the linear velocity appeared very nearly proportional to the radius (in accordance with Fig. 4.3), and outside the core the velocity closely checked the values of Eq. (9.37). The vortex span was somewhat greater than predicted in Sect. 11.6 but remained constant with α.

The diameter of the vortex core (for which no satisfactory theory has been devised) increased slightly with the circulation. The center of the core for this particular case revolved at over 18,000 rpm.

The high speed of the vortex core produces interesting and beautiful effects in air under special conditions. The high speed implies a low pressure, and for air we then get a low temperature. Since the amount of moisture the air can hold decreases with falling temperature, vortices in humid air becomes visible, appearing somewhat like steam in air, as the moisture condenses out. This phenomenon is a familiar sight at the wing tips and landing-flap terminations and, more frequently, streaming helically behind propeller blade tips at take-off.

9.7. Induced Drag Coefficient. We may now examine the effect of the downwash on drag for the simplest case—that of elliptic loading.

For the finite wing, Eq. (4.25) becomes

$$L = \rho V \int_{-b/2}^{b/2} \Gamma \, dy \tag{9.38}$$

and for elliptic loading in accordance with Eq. (9.33) we get

$$L = \rho V \Gamma_0 \int_{-b/2}^{b/2} \sqrt{1 - \left(\frac{2y}{b}\right)^2} \, dy \tag{9.39}$$

Making the substitution that $y = (b/2) \sin \theta$, $dy = (b/2) \cos \theta \, d\theta$ and finding new limits we have

$$L = \rho V \Gamma_0 \int_{-\pi/2}^{\pi/2} \sqrt{1 - \sin^2 \theta} \, \frac{b}{2} \cos \theta \, d\theta \tag{9.40}$$

Integrating and solving for Γ_0, we have

$$\Gamma_0 = \frac{4L}{\rho \pi V b} \tag{9.41}$$

Substituting into Eq. (9.36),

$$w = \frac{\Gamma_0}{2b} = \frac{2L}{\rho \pi V b^2} \tag{9.42}$$

and it is seen that the effect of downwash is to deflect the airstream through an angle

$$\alpha_i = \frac{w}{V} \tag{9.43}$$

as is shown in Fig. 9.18.   The changes in lift and velocity are small and may be neglected, but the effect on drag may be large.

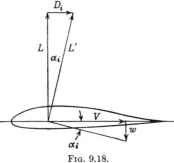

Considering the dragwise component of the lift D_i, we have

$$D_i = L \tan \alpha_i = L \left(\frac{w}{V} \right) \tag{9.44}$$

and from Eq. (9.43) this becomes

$$D_i = \frac{L^2}{(\rho/2) V^2 \pi b^2} \tag{9.45}$$

Fig. 9.18.

Multiplying both sides of the equation by V and dividing by 550 to get horsepower we have, since the lift must equal the weight W in steady unaccelerated flight,

$$\text{hp}_i = \frac{(W/b)^2 V}{550 \pi q} \tag{9.46}$$

where q = dynamic pressure $(\rho/2) V^2$. Putting V in miles per hour and letting $\sigma = \rho/\rho_0$, where ρ_0 = density at standard conditions, the final form is

$$\text{hp}_i = \frac{(W/b)^2}{3.0 \sigma V_{\text{mph}}} \tag{9.47}$$

Equation (9.47) demonstrates that the horsepower required to overcome induced drag is a function of the *span loading*, the *velocity*, and the *density* and gives a clearer insight into the problem than the usual

$$C_{Di} = \frac{C_L^2}{\pi \, \text{A.R.}} \tag{9.48}$$

which may be found from Eq. (9.45). Equation (9.48) seems to indicate that induced drag is a function of aspect ratio, where, more accurately, only the drag *coefficient* varies with aspect ratio, and the area changes that occur with changing aspect ratio and *constant* span are such that the induced drag remains constant with constant weight and speed.

It should be noted that Eqs. (9.45), (9.47), and (9.48) were derived for elliptic loading only.

PROBLEMS

9.1. If a wing of 6 ft chord has a local lift coefficient of 0.6 and is operating in a 100-mph airstream, what is the local circulation?

9.2. Prove that the untwisted elliptic wing gives elliptic distribution of circulation.

9.3. What is the downwash in feet per second at the wing centerline of a wing with uniform loading if its lift is 10,000 lb, span 30 ft, $V = 200$ mph, and the air standard sea-level density?

9.4. Find the induced velocity 4 ft from a doubly infinite vortex of strength $\Gamma = 30$ sq ft per sec.

REFERENCES

9.1. F. W. Lanchester, "Aerodynamics," Constable & Co., Ltd., London, 1918.

9.2. L. Prandtl, Applications of Modern Hydrodynamics to Aeronautics, *TR* 116, 1921.

9.3. N. A. V. Piercy, On the Vortex Pair Quickly Formed by Some Airfoils, *JRAS*, October, 1923, p. 489.

9.4. S. Goldstein, Modern Developments in Fluid Dynamics, Vol. I, p. 69, Oxford University Press, New York, 1938.

9.5. N. A. V. Piercy, "Aerodynamics," p. 235, English University Press, London.

CHAPTER 10

THE MONOPLANE WING

10.1. The Fundamental Wing Equation. Continuing our conception of the wing as represented by horseshoe vortices, we turn now to the consideration of each section of the wing as affected by the entire velocity field.

It is seen immediately that the local angle of attack of a section of an untwisted wing of finite span is not the geometrical angle between the section chord and the free stream, but that angle diminished by the local downwash angle w/V. Further, since w/V varies across the span of most finite wings, the local angle of attack varies too, and it is evident that any airfoil characteristic that is a function of α [such as $c_{l\,\text{max}}$, $(L/D)_{\text{max}}$, etc.] probably will not occur simultaneously across the span. Thus a finite wing is usually "worse" than its airfoil section. The maximum lift of the wing, for instance, is frequently 20 per cent less than the airfoil maximum lift coefficient.

In the following pages we shall apply the Biot-Savart law to several wing planforms and determine how serious the variation of angle of attack along the span is in specific cases. In addition we shall examine design parameters which minimize this effect.

Expressing the local or section angle of attack mathematically, we have

$$\alpha_0(y) = \alpha(y) - \frac{w(y)}{V} \tag{10.1}$$

where $\alpha(y)$ = geometric angle of attack at sta. y
$\alpha_0(y)$ = effective (two-dimensional) angle of attack at sta. y
$w(y)$ = downwash at sta. y
The value of w at y_1 as shown previously is

$$w(y_1) = \frac{1}{4\pi} \int_{-b/2}^{b/2} \frac{\dfrac{\partial \Gamma}{\partial y}\, dy}{y_1 - y} \tag{10.2}$$

According to the method of Sect. 2.7 we may express a distribution of circulation symmetrical about a centerline as a sine series, keeping only odd values of n. First, however, we must describe the wing in terms of trigonometric functions. Let $y = -s \cos \theta$, where s = the wing semi-span. Then $dy = s \sin \theta\, d\theta$. The arbitrary series for the circulation becomes

$$\Gamma = 4sV \sum_1^\infty A_n \sin n\theta*$$ (10.3)

and

$$\frac{\partial \Gamma}{\partial \theta} = 4sV \sum_1^\infty nAn \cos n\theta$$ (10.4)

At a particular point on the wing θ_1 (corresponding to y_1)

$$w(\theta_1) = \frac{1}{4\pi} \int_0^\pi \frac{\frac{\partial \Gamma}{\partial \theta}\, d\theta}{y_1 - y}$$ (10.5)

$$w(\theta_1) = \frac{V}{\pi} \int_0^\pi \frac{\sum_1^\infty nA_n \cos n\theta}{\cos \theta_1 - \cos \theta}\, d\theta$$ (10.6)

By the method of Appendix 3, it may be shown that

$$\int_0^\pi \frac{\cos n\theta}{\cos \theta_1 - \cos \theta}\, d\theta = \frac{\pi \sin n\theta_1}{\sin \theta_1}$$ (10.7)

where θ_1 is a constant. Hence

$$w(\theta_1) = V \sum_1^\infty nAn \frac{\sin n\theta_1}{\sin \theta_1}$$ (10.8)

Since θ_1 was perfectly arbitrary, Eq. (10.8) holds for a general point θ so that

$$w \sin \theta = V \sum_1^\infty nAn \sin n\theta$$ (10.9)

As $\alpha_0 = c_l/a_0$ and [Eq. (4.26)] $c_l = 2\Gamma/cV$, Eq. (10.1) becomes [using Eqs. (10.4) and (10.9) and letting $\mu = a_0c/8s$]

$$\sum_1^\infty A_n \sin n\theta(\mu n + \sin \theta) = \mu \alpha \sin \theta$$ (10.10)

* These coefficients A_n have no connection with those used in Chaps. 7 and 8, of course.

Equation (10.10) is the fundamental equation for determining the coefficients A_n for a monoplane wing. Its use will be discussed more fully later.

10.2. Lift and Drag. We now substitute the value of from Eq. (10.3) into the equation of wing lift:

$$L = \int_{-s}^{s} \rho V \Gamma \, dy \tag{10.11}$$

$$L = 4\rho s^2 V^2 \int_0^\pi \sum_1^\infty A_n \sin n\theta \sin \theta \, d\theta \tag{10.12}$$

But $\int_0^\pi \sin n\theta \sin \theta \, d\theta = 0$ for all values of n except $n = 1$, at which time (integrating) it becomes $\pi/2$. Equation (10.12) hence reduces to

$$L = \frac{\rho}{2} V^2 \cdot 4\pi s^2 \cdot A_1 \tag{10.13}$$

and

$$C_L = \frac{4\pi s^2}{S} A_1 \tag{10.14}$$

Hence

$$A_1 = \frac{S}{4\pi s^2} C_L = \frac{C_L}{\pi \text{ A.R.}} \tag{10.15}$$

where A.R. = wing aspect ratio = $\dfrac{b^2}{S} = \dfrac{4s^2}{S}$.

The induced drag

$$D_i = \frac{w}{V} L$$

or

$$D_i = \int_{-s}^{s} \rho V \Gamma \frac{w}{V} \, dy$$

Substituting from Eqs. (10.7) and (10.10), we have

$$D_i = 4\rho s^2 V^2 \int_0^\pi \sum_1^\infty n A_n^2 \sin^2 n\theta \, d\theta$$

$$= 2\rho s^2 V^2 \pi \sum_1^\infty n A_n^2 \tag{10.16}$$

where, as will be shown later, the constants A_n may be determined by the wing geometry and Eq. (10.10).

10.3. Minimum Induced Drag. Of considerable interest is the spanwise distribution of circulation that yields minimum induced drag.

Since in Eq. (10.16) ρ, s, and V are necessary, we turn our attention to the expression $\sum_1^\infty nA_n{}^2$. The minimum condition for the series is, of course, $n = 0$, but it was shown in Sect. 10.2 that $n = 1$ was necessary for lift. Since the terms are squared (and minus signs may hence be neglected), two terms will always have a larger value than one, and we conclude that the minimum induced drag is

$$D_{i,\text{min}} = 2\rho s^2 V^2 \pi A_1{}^2 \tag{10.17}$$

For the minimum induced drag condition, Eq. (10.3) becomes

$$\Gamma = 4sVA_1 \sin\theta \tag{10.18}$$

Since $y = -s\cos\theta$, $\cos\theta = -y/s$ and $\sin\theta = \sqrt{1 - (y/s)^2}$, Eq. (10.18) becomes

$$\frac{\Gamma^2}{16s^2V^2A_1{}^2} + \frac{y^2}{s^2} = 1 \tag{10.19}$$

Equation (10.19) is that of an ellipse and demonstrates that the minimum induced drag is realized when the spanwise distribution of circulation is elliptic. This desirable condition may be attained by having a wing of constant profile and elliptic distribution of chord along the span with zero twist. If the 50 per cent chord line is straight, the planform of the wing will be an ellipse, a variety of elliptic chord wings being possible with arbitrary mean lines.

Equation (10.17) may be rewritten as

$$D_{i,\text{min}} = \frac{\rho}{2} SV^2 \frac{C_L{}^2}{\pi \text{ A.R.}} \tag{10.20}$$

or in horsepower form as in Eq. (9.47).

Before continuing, it should be emphasized that this section and its conclusions are valid only for wings that may be simulated by simple horseshoe vortex systems—systems such that the downwash at the wing is due only to the trailing vortices. This is ordinarily the case. But special situations such as low aspect ratios (when the chordwise distribution of vorticity becomes important), flying near the ground (when the image airplane influences the flow near the real airplane), or flying near another airplane may easily result in *less* than the above indicated minimum drag.

10.4. Aspect-ratio Corrections for Induced Flow: Angle of Attack. It has been shown that the trailing vortices impart a downward velocity to the airstream at the wing and hence that the lift of any section of the

wing is less than the two-dimensional lift for the same geometric angle
of attack.   The general monoplane equation

$$\sum_{1}^{\infty} A_n \sin n\theta(\mu n + \sin \theta) = \mu\alpha \sin \theta \qquad (10.21)$$

may be used to determine the coefficients A_n and in particular the coefficient proportional to lift A_1.   Since the general equation contains the
angle of attack α and determines the lift by determining A_1, it also determines the slope of the wing lift curve a.   From Eq. (10.15)

$$C_L = \pi A_1 \,(\text{A.R.}) = \alpha a \qquad (10.22)$$

where α = aerodynamic angle of attack $\equiv$ angle of attack measured from
zero lift

Hence

$$a = \frac{\pi A_1 \,(\text{A.R.})}{\alpha} \qquad (10.23)$$

Now a certain value of the lift coefficient may be expressed as a section (two-dimensional) slope times an effective (two-dimensional) angle
of attack, as

$$C_L = \alpha_0 a_0 \qquad (10.24)$$

or as a wing lift curve slope times a wing angle of attack as in Eq. (10.22).
Hence

$$\alpha = \frac{C_L}{a} \qquad \text{and} \qquad \alpha_0 = \frac{C_L}{a_0} \qquad (10.25)$$

and

$$\alpha - \alpha_0 = C_L \left(\frac{1}{a} - \frac{1}{a_0}\right) \qquad (10.26)$$

Substituting $A_1 = C_L/\pi \text{ A.R.}$ and $a = C_L/\alpha$, we get

$$\alpha - \alpha_0 = \frac{C_L}{\pi \text{ A.R.}} \left(\frac{\alpha}{A_1} - \frac{\pi \text{ A.R.}}{a_0}\right)$$

and setting

$$\frac{\alpha}{A_1} - \frac{\pi \text{ A.R.}}{a_0} = 1 + \tau \qquad (10.27)$$

we get

$$\alpha = \alpha_0 + \frac{C_L}{\pi \text{ A.R.}} (1 + \tau) \qquad (10.28)$$

It is seen that τ is a small positive factor that increases the induced
angle over that for the minimum case of elliptic loading to make allow-

ances for deviations from elliptic loading. We shall evaluate it for specific wing planforms later.

10.5. Aspect-ratio Corrections for Induced Flow: Drag. The relation for the induced drag of a wing [Eq. (10.16)] may be simplified if we write

$$1 + \delta = \sum_{1}^{\infty} n \frac{A_n^2}{A_1^2} \tag{10.29}$$

where δ is some small positive quantity. It then becomes

$$D_i = 2\rho s^2 V^2 \pi A_1^2 (1 + \delta)$$

and the induced drag for any wing is now related to that for an elliptic wing by δ. Substituting from Eq. (10.14) and dividing both sides by $(\rho/2)V^2S$, we get

$$C_{Di} = \frac{C_L^2}{\pi \text{ A.R.}} (1 + \delta) \tag{10.30}$$

and the total wing drag is then

$$C_D = C_{D0} + \frac{C_L^2}{\pi \text{ A.R.}} (1 + \delta)$$

where C_{D0} = wing profile drag.

It is seen that δ compares with τ [of Eq. (10.28)], and experience shows us it represents a small additive correction to the induced drag for elliptic loading to allow for deviations from that minimum condition.

Actually in practice the values of τ and δ are rarely computed by themselves. In general the facts presented above and in Sect. 10.12 to the effect that δ is smallest for taper ratios of around $2\frac{1}{2}$ is used as a background for employing that taper, small variations arising when structural or dimensional conditions become important.

The mechanism usually employed instead of the δ form is to lump all drag other than that at $C_L = 0$ into the induced drag (calling it an *effective induced drag*) by means of an efficiency factor e which is less than 1.0. Thus

$$C_{D,\text{total}} = C_{D,\text{min}} + \frac{C_L^2}{\pi(eb)^2/S} \tag{10.31}$$

The shortened span eb is called the *effective span*, and $(eb)^2/S$ is called the *effective aspect ratio*. The factor e accounts for (1) the increase of fuselage drag with C_L; (2) the increase of wing profile drag with C_L; (3) the increase of tail drag with C_L; (4) the increase of drag due to nonelliptic span loading; and (5) other miscellaneous losses that vary with C_L.

The use of e is adequately covered in many textbooks on practical aerodynamics and will not further be discussed here except to note that the almost universal use of taper ratios near $2\frac{1}{2}$ to 1 springs directly from theoretical considerations. Indeed, at the higher aspect ratios (see Fig. 10.10) it is particularly important to employ elliptic planforms or the proper taper ratio to avoid seriously large increases of induced drag.

10.6. Corrections for Change of Aspect Ratio. When the wing coefficients are known for one aspect ratio, $(\ )_1$, Eqs. (10.28) and (10.31) may be rearranged to solve for the coefficients at a second aspect ratio, $(\ )_2$.

$$\alpha_1 = \alpha_0 + \frac{C_L}{\pi \text{ A.R.}} (1 + \tau_1)$$

$$\alpha_2 = \alpha_0 + \frac{C_L}{\pi \text{ A.R.}} (1 + \tau_2)$$

and

$$\alpha_2 = \alpha_1 + \frac{C_L}{\pi} \left(\frac{1 + \tau_2}{\text{A.R.}_2} - \frac{1 + \tau_1}{\text{A.R.}_1} \right) \tag{10.32}$$

Similarly

$$C_{L2} = C_{D1} + \frac{C_L^2}{\pi} \left(\frac{1 + \delta_2}{\text{A.R.}_2} - \frac{1 + \delta_1}{\text{A.R.}_1} \right) \tag{10.33}$$

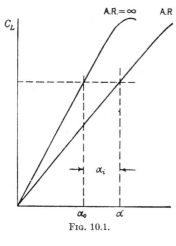

FIG. 10.1.

These equations are less useful than the empirical airplane efficiency factors described in Sect. 10.5.

10.7. Slope of a Three-dimensional Elliptic-wing Lift Curve. Using the relation developed above, we may now derive the theoretical slope of the lift curve for an elliptic wing of any aspect ratio. From Eq. (10.28) and Fig. 10.1,

$$\alpha = \alpha_0 + \frac{C_L}{\pi \text{ A.R.}}$$

and we see that C_L/π A.R. is the induced angle α_i of an elliptic wing as well as the coefficient of the first term of the circulation series. That is

$$\alpha_i = \frac{C_L}{\pi \text{ A.R.}} = A_1 = \frac{w_0}{V}$$

where w_0 = downwash at an elliptic wing; it is constant across the span.

Since

$$\alpha = \frac{C_L}{a}$$

where a = lift curve slope $\frac{dC_L}{d\alpha}$ per radian

$$\frac{C_L}{a} = \frac{C_L}{a_0} + \frac{C_L}{\pi \, \text{A.R.}}$$

Differentiating with respect to C_L, we have

$$a = \frac{\pi a_0 \, \text{A.R.}}{a_0 + \pi \, \text{A.R.}} \tag{10.34}$$

Assuming the infinite-aspect-ratio lift curve slope a_0 for a thin wing to be 2π per radian,

$$a = a_0 \frac{\text{A.R.}}{\text{A.R.} + 2} = 2\pi \frac{\text{A.R.}}{\text{A.R.} + 2} \tag{10.35}$$

Practicing aerodynamicists frequently replace the constant in the denominator of Eq. (10.35) by 2.5 or even 3.0 since their experience has shown that such a change improves the correlation with experimental results.

TABLE 10.1. LIFT CURVE FOR ELLIPTIC LOADING

A.R.	∞	10	8	6	4
a (per radian)..........	6.28	5.236	5.027	4.712	4.188
a (per deg).............	0.1097	0.0914	0.0877	0.0823	0.073

10.8. The Edge Correction. Jones (Ref. 10.1) points out that a correction of (Eq. 10.34) for three-dimensional flow is quite in order. For instance, consider that the maximum, or "edge," velocity around an infinite circular cylinder is $2V$, while that around a sphere is $1.59V$ (V is the free-stream velocity). The conception is that three-dimensionally the air has more "places to go" as it passes over a three-dimensional body and hence develops less velocity increment (see Fig. 10.2).

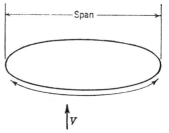

FIG. 10.2. Planview of elliptic wing, showing increase of "escape area" for three-dimensional flow.

Applied to an elliptic wing, this concept indicates that the edge velocity is, on the average, decreased by the factor:

$$E = \frac{\text{wing semi-perimeter}}{\text{wing span}}$$

and since the edge velocity determines the circulation through the Kutta condition, the lift is similarly decreased.

The correction may be put into the lift curve slope formula as follows: From thin-airfoil theory

$$C_L = 2\pi(\alpha - \alpha_i) = 2\pi\left(\alpha - \frac{C_L}{\pi \text{ A.R.}}\right)$$

which with the edge correction becomes

$$C_L = 2\pi\alpha \frac{\text{A.R.}}{E \text{ A.R.} + 2}$$

Starting anew with Eq. (10.34) and including the effects of airfoil thickness from Eq. (5.65) and the edge correction, the final complete equa-

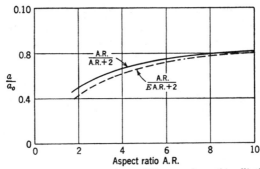

FIG. 10.3.   Effect of edge correction on lift curve slope, thin elliptic wings.

tion for the slope of the lift curve of an elliptic wing becomes

$$a = 2\pi k \frac{\text{A.R.}}{E \text{ A.R.} + 2k} \qquad (10.36)$$

where
$$k = \frac{1 + \epsilon}{1 + \epsilon^2}$$

$$\epsilon = \frac{4}{3\sqrt{3}} \times \text{airfoil thickness ratio}$$

as in Eq. (5.65).

The relative size of the edge correction is shown in Fig. 10.3, and the theoretical slope of the lift curve for a number of elliptic wings is given in Fig. 10.4.

10.9. Rectangular Wings. The values of τ and δ are of interest for all planforms in order to find how serious a deviation from elliptic loading will be. The simplest case to examine is that of a rectangular wing. (Tapered wings are considered in another section.)

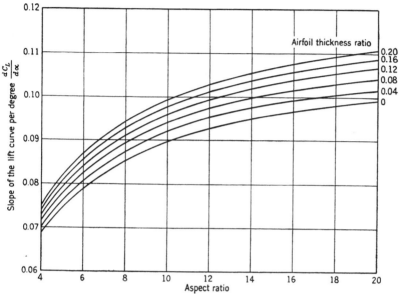

Fig. 10.4. Theoretical values of the slope of the lift curve for elliptic wings, edge correction included.

The general expression for circulation

$$\Gamma = 4sV \sum_{1}^{\infty} A_n \sin n\theta \tag{10.37}$$

will, of course, be applicable to the rectangular wing, and we shall use the fundamental monoplane equation

$$\sum_{1}^{\infty} A_n \sin n\theta(\mu n + \sin \theta) = \mu \alpha \sin \theta$$

to find the values of A_n and then Eqs. (10.27) and (10.29) to find τ and δ for a specific case. The span stations will as usual be expressed in the form $y = -s \cos \theta$, and the term μ, which for elliptic or tapered wings varies along the span, becomes a constant for the rectangular wing as follows:

$$\mu = \frac{a_0}{4 \text{ A.R.}}$$

The successive coefficients A_1, A_3, A_5, A_7 decrease rapidly in size so that ample accuracy is obtained by keeping four only. The fundamental monoplane equation becomes

$$A_1 \sin \theta(\mu + \sin \theta) + A_3 \sin 3\theta(3\mu + \sin \theta) + A_5 \sin 5\theta(5\mu + \sin \theta)$$
$$+ A_7 \sin 7\theta(7\mu + \sin \theta) = \mu\alpha \sin \theta \quad (10.38)$$

and by selecting four span stations at which we should like to find the circulation (four values of θ, that is) we shall obtain four equations in four unknowns and hence a solution. Nearly any well-distributed span stations would be satisfactory, a slight advantage accruing if we select $\theta = 22.5, 45.0, 67.5$, and $90°$, as through the mechanism of $y = -s \cos \theta$ we then get points displaced toward the wing tip, where the greatest variation occurs.

For example, consider that the values for A_n are wanted for a rectangular wing of aspect ratio of 6. Then

$$\mu = \frac{2\pi}{4 \text{ A.R.}} = 0.262$$

Substituting the four angles in Eq. (10.38), we have

$$0.247 \frac{A_1}{\mu\alpha} + 1.080 \frac{A_3}{\mu\alpha} + 1.564 \frac{A_5}{\mu\alpha} + 0.862 \frac{A_7}{\mu\alpha} = 0.383 \quad (10.39)$$

$$0.685 \frac{A_1}{\mu\alpha} + 1.055 \frac{A_3}{\mu\alpha} - 1.421 \frac{A_5}{\mu\alpha} - 1.797 \frac{A_7}{\mu\alpha} = 0.707 \quad (10.40)$$

$$1.097 \frac{A_1}{\mu\alpha} - 0.654 \frac{A_3}{\mu\alpha} - 0.854 \frac{A_5}{\mu\alpha} + 2.545 \frac{A_7}{\mu\alpha} = 0.924 \quad (10.41)$$

$$1.262 \frac{A_1}{\mu\alpha} - 1.786 \frac{A_3}{\mu\alpha} + 2.311 \frac{A_5}{\mu\alpha} - 2.835 \frac{A_7}{\mu\alpha} = 1.000 \quad (10.42)$$

A good method of solving Eqs. (10.39) through (10.42) is by the use of determinants. However, as these simultaneous equations form a fourth-order determinant, reduction according to proper laws is first in order. The solution finally becomes

$$\frac{A_1}{\mu\alpha} = 0.917, \quad \frac{A_3}{\mu\alpha} = 0.1108, \quad \frac{A_5}{\mu\alpha} = 0.0218, \quad \frac{A_7}{\mu\alpha} = 0.00373$$

and we confirm, fortunately, that the terms rapidly decrease in size so that a four-point solution is adequate.

From Eqs. (10.27) and (10.29) and the above,

$$\tau = \frac{1}{\mu}\left(\frac{\mu\alpha}{A_1} - \frac{\pi}{4}\right) - 1$$

$$= \frac{1}{0.262}\left(\frac{1}{0.917} - \frac{\pi}{4}\right) - 1$$

$$= 0.1615$$

$$\delta = \sum_1^\infty n\frac{A_n^2}{A_1^2} - 1 \tag{10.43}$$

$$= \frac{A_1^2 + A_3^2 + A_5^2 + A_7^2}{A_1^2} - 1$$

$$= \frac{(0.917)^2 + 3(0.1108)^2 + 5(0.0218)^2 + 7(0.00373)^2}{(0.917)^2} - 1$$

$$= 0.045$$

which means that [in accordance with Eqs. (10.28) and (10.30)] the induced angles and induced drag for a rectangular wing of A.R. = 6.0 should be increased 16.15 and 4.5 per cent, respectively. These corrections are not as serious as they

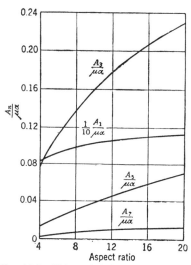

FIG. 10.5. Values of $A_n/\mu\alpha$ for rectangular wings.

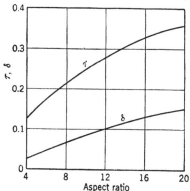

FIG. 10.6. Values of τ and δ for rectangular wings.

appear since the induced drag and angles may be only a fraction of the total values.

Other values of $A_n/\mu\alpha$ are shown in Fig. 10.5 and of τ and δ in Fig. 10.6.

The spanwise distribution of circulation for a rectangular wing may now be found by substituting the values of A_n into the expanded expression for Γ:

$$\Gamma = 4sV(A_1 \sin \theta + A_3 \sin 3\theta + A_5 \sin 5\theta + A_7 \sin 7\theta)$$

The spanwise distribution of circulation obtained by this method is shown for a wing of A.R. = 6.28 in Fig. 10.7. The broken line is the mean between the chord distribution and a semi-ellipse having the same area as the wing. The significance of this approximation is discussed in Chap. 12.

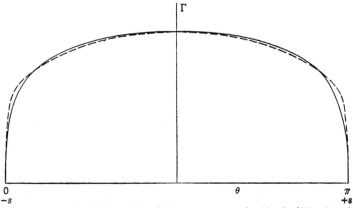

Fig. 10.7. Spanwise distribution of circulation over rectangular wing by lifting-line theory (solid line) and an approximate curve (broken).

10.10. Slope of Rectangular-wing Lift Curve. The slope of the lift curve for rectangular or tapered wings follows from Eq. (10.36) by including the effect of τ on the induced angle. The final result is

$$a = 2\pi k \frac{\text{A.R.}}{E \text{ A.R.} + 2k(1 + \tau)} \tag{10.44}$$

where k and ϵ are the same as in (Eq. 10.36), and τ must be determined by the method of Sect. 10.9. Values of Eq. (10.44) are plotted in Fig. 10.8 for a number of aspect and thickness ratios.

10.11. Rectangular-wing Pitching Moment. From the usual conception of the existence of an aerodynamic center about which the moment coefficient is constant for reasonable values of c_l (or α), we may directly infer that the complete wing coefficient about the aerodynamic center should be the same as that of the airfoil, for the variation of the effective angle of attack along the span should have no effect on c_{mac}. That is,

$$C_{mae} = c_{mac} \tag{10.45}$$

Hence a rectangular wing should have the same moment coefficient as its airfoil section when they are examined at equal lift coefficients.

In practice, according to Bradfield (Ref. 10.2), Eq. (10.45) agrees best for aspect ratios near 6.0. Below that the aerodynamic center moves

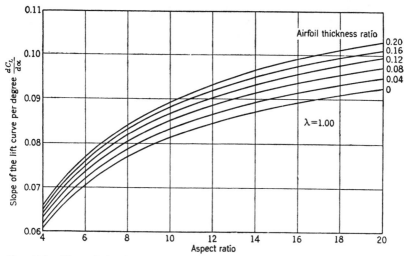

Fig. 10.8. Theoretical values for the slope of the lift curve for rectangular wings, edge correction included.

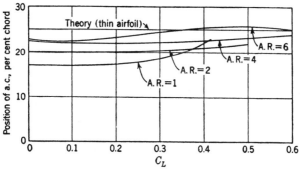

Fig. 10.9. Comparison of theoretical and experimental determination of aerodynamic-center location, rectangular wings.

forward with decreasing aspect ratio. The reason for this action can be intuitively derived by considering a factor not included in the discussion above:

The trailing vortices induce a downward velocity field (Fig. 9.1) that increases toward the wing trailing edge so that, effectively, the cambered airfoil is in a cambered flow and the relative curvature between the two is hence reduced.   Since the moment about the aerodynamic center depends on this relative curvature, its value is reduced both by increasing lift (which increases the curvature for a given aspect ratio) and by decreased aspect ratio (which increases the curvature for a given lift).

10.12. Tapered Wings.   We may now apply the general monoplane equation to the case of a tapered wing, following a method used by Glauert in Ref. 10.3.   First let the ratio of the tip chord to the root chord be λ.   The local chord will be

$$c = c_0[1 - (1 - \lambda) \cos \theta]$$

where c_0 = center-section chord.

The parameter μ becomes

$$\mu = \mu_0[1 - (1 - \lambda) \cos \theta] \tag{10.46}$$

and

$$\mu_0 = \frac{a_0 c_0}{8s} \tag{10.47}$$

where a_0 = slope of the lift curve for infinite aspect ratio.

Using the above relations, the wing area is

$$S = sc_0(1 + \lambda)$$

and the aspect ratio is

$$\text{A.R.} = \frac{a_0}{2\mu_0(1 + \lambda)} \tag{10.48}$$

which is used for finding μ_0.

Proceeding as in Eq. (10.38), we replace μ by $\mu_0[1 - (1 - \lambda) \cos \theta]$ and, again selecting values of θ of 22.5, 45, 67.5, and 90°, get four simultaneous equations, from which the coefficients A_1/α, A_3/α, A_5/α, and A_7/α may be found.   Equations (10.27) and (10.29) may then be used to solve for δ and τ, enabling us to compare the aerodynamic performance of tapered wings with those of elliptic and rectangular planforms.

For example, for aspect ratio 6.0, taper ratio 0.75 ($\lambda = 0.75$), μ_0 becomes 0.299.   The simultaneous equations are

$$2.665 \frac{A_1}{\alpha} + 11.25 \frac{A_3}{\alpha} + 16.07 \frac{A_5}{\alpha} + 8.67 \frac{A_7}{\alpha} = 1.00$$

$$3.873 \frac{A_1}{\alpha} + 5.874 \frac{A_3}{\alpha} - 7.88 \frac{A_5}{\alpha} - 9.88 \frac{A_7}{\alpha} = 1.00$$

$$4.420 \frac{A_1}{\alpha} - 2.661 \frac{A_3}{\alpha} - 3.490 \frac{A_5}{\alpha} + 10.42 \frac{A_7}{\alpha} = 1.00$$

$$4.342 \frac{A_1}{\alpha} - 6.28 \frac{A_3}{\alpha} + 8.34 \frac{A_5}{\alpha} - 10.34 \frac{A_7}{\alpha} = 1.00$$

Solving, $A_1/\alpha = 0.242$, $A_3/\alpha = 0.020$, $A_5/\alpha = 0.008$, and $A_7/\alpha = 0.0$. By Eq. (10.29)

$$\delta = \frac{(0.242)^2 + 3(0.020)^2 + 5(0.008)^2 + 7(0)^2}{(0.242)^2} - 1 = 0.025$$

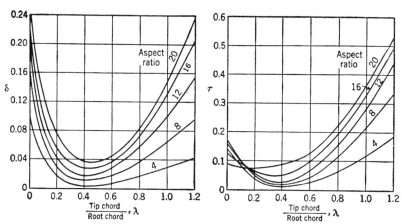

Fig. 10.10.   Values of δ and τ for various aspect and taper ratios.

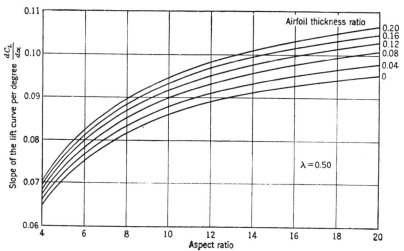

Fig. 10.11.   Theoretical values for the slope of the lift curve for wings with 2 to 1 taper, edge correction included.

We would therefore expect the induced drag of a wing of A.R. = 6 with a taper ratio of 0.75 to be 2.5 per cent above that of an elliptic wing of the same aspect ratio.

Additional calculations for many other aspect ratios yield the data shown in Fig. 10.10. Of some interest is the fact that a taper ratio of about 2.7:1 yields the nearest approach to elliptic loading drag for an aspect ratio of 4, while about 2.5:1 is best for an aspect ratio of 12.0.

Theoretical lift curve slopes for wings with a taper ratio of 0.5 are plotted in Fig. 10.11 as obtained from Eq. (10.44). Other taper ratios may be considered through Eq. (10.44), directly.

The usually good agreement between the values of Eq. (10.44) and experiment may be improved when the section lift curve slope is known by replacing k by $a_0/2\pi$, where a_0 is the experimentally determined section lift curve slope.

The effect of taper on the spanwise load distribution and on the stall pattern is discussed in Chap. 12.

PROBLEMS

10.1. Show that τ and δ in Sects. 10.4 and 10.5 are for rectangular wings and not for uniformly loaded wings.

10.2. Calculate check values for Figs. 10.5 and 10.6.

10.3. Compare the total induced drag of two rectangular wings with (a) one far behind the other and (b) with the leading edge of the rear touching the trailing edge of the front one.

10.4. Find the horsepower required to overcome induced drag of a modified B-29, span 140 ft, $W = 120,000$ lb, wing area 2,000 sq ft with (a) elliptic wing, (b) rectangular wing, (c) wing with 2:1 taper. All wings have same aspect ratio. $V_{ind} = 150$ mph, 30,000 ft altitude, standard air.

10.5. Prove that maximum L/D is obtained when the induced drag equals the profile drag.

REFERENCES

10.1. R. T. Jones, Correction of the Lifting-line Theory for the Effect of the Chord, *TN* 817, 1941.

10.2. F. B. Bradfield, Center of Pressure Travel of Symmetrical Sections at Small Angles of Incidence, *R & M* 1294, 1929.

10.3. H. Glauert, The Characteristics of a Tapered and Twisted Wing with Sweepback, *R & M* 1226, 1928.

CHAPTER 11

THE FIELD ABOUT THE WING

11.1. The Field near the Horseshoe Vortex. According to the Biot-Savart law a field of induced velocities exists around the horseshoe vortex system with which we simulate the wing and its trailing vortices. We may now examine this field and, using the Biot-Savart law obtain first-approximation values for the various induced velocities. The results have practical value for an understanding of pitot-static-tube position error, span load distribution, wind-tunnel boundary corrections, and ground effect.

Two types of span loading (uniform and elliptic) are simply solved for their respective fields and will be considered in the following pages. It will develop that the expedient of using simple vortices to represent the wing yields reasonable values for all but the case of downwash at the tail. Here additional factors that allow for the downward displacement of the trailing vortices must be employed for best results.

The axes previously used will suffice. The x axis is in the plane of symmetry positive upstream, the y axis coincides with the bound wing vortex positive to starboard, and the z axis extends downward at the intersection of the x and y axes.

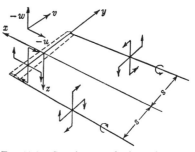

FIG. 11.1. Set of axes and a horseshoe vortex showing the velocity components of each segment.

It will be seen that this is an orthogonal right-hand system.

In order to visualize the origin of the flow field about the wing, let us assign subscripts to the velocity components u, v, and w for each of the three vortices: for the bound vortex ( )$_b$; for the starboard vortex ( )$_s$; for the port vortex ( )$_p$. We see that the total induced flow at (x, y, z) will be (see Fig. 11.1)

$$u = u_b + u_s + u_p \tag{11.1}$$
$$v = v_b + v_s + v_p \tag{11.2}$$

and

$$w = w_b + w_s + w_p \tag{11.3}$$

Some simplification occurs since from the normality of the bound and trailing vortices the longitudinal velocity component u depends only on the bound vortex so that we may write

$$u = u_b \tag{11.4}$$

and the lateral velocity v depends only on the trailing vortices s and p so that

$$v = v_s + v_p \tag{11.5}$$

The vertical velocity w is composed of all three components.

In the following sections some liberty has been taken with exact mathematical nomenclature in order to simplify the expressions as much as possible. First of all, the minus signs have been left off terms which through squaring become plus; second, absolute-value bars have not been used as frequently as exact rigor would dictate. It is believed that these procedures tend toward clarity and brevity with no appreciable loss in exactitude.

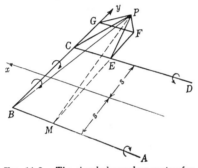

FIG. 11.2. The simple horseshoe vortex for uniform loading.

11.2. Uniform Loading. When uniform loading is assumed, the circulation has a constant strength Γ which is also the strength of both trailing vortices. Consider a general point P as shown in Fig. 11.2. The induced velocity at P due to the bound vortex will be normal to plane PBG and will have a value

$$q_b = \frac{\Gamma}{4\pi(PG)} (\cos \angle PBC + \cos \angle PCB) \tag{11.6}$$

$$= \frac{\Gamma}{4\pi \sqrt{x^2 + z^2}} \left[\frac{y + s}{\sqrt{(y + s)^2 + x^2 + z^2}} - \frac{y - s}{\sqrt{(y - s)^2 + x^2 + z^2}} \right] \tag{11.7}$$

The components of the velocity q_b are

$$u_b = q_b \sin \angle PGF = q_b \cdot \frac{PF}{PG} \tag{11.8}$$

$$= \frac{+\Gamma z}{4\pi(x^2 + z^2)} \left[\frac{y + s}{\sqrt{(y + s)^2 + x^2 + z^2}} - \frac{y - s}{\sqrt{(y - s)^2 + x^2 + z^2}} \right] \tag{11.9}$$

$$v_b = 0$$

$$w_b = q_b \cos \angle PGF = q_b \frac{GF}{PG}$$

$$= \frac{-\Gamma x}{4\pi(x^2 + z^2)} \left[\frac{y + s}{\sqrt{(y + s)^2 + x^2 + z^2}} - \frac{y - s}{\sqrt{(y - s)^2 + x^2 + z^2}} \right] \tag{11.10}$$

The induced velocity at P due to the starboard vortex s is

$$q_s = \frac{\Gamma}{4\pi \, PE} \left(\cos \angle PDC + \cos \angle PCE \right) \tag{11.11}$$

$$= \frac{\Gamma}{4\pi \sqrt{z^2 + (y - s)^2}} \left[1 + \frac{-x}{\sqrt{(y - s)^2 + x^2 + z^2}} \right] \tag{11.12}$$

$$u_s = 0$$

$$v_s = q_s \sin \angle PEF$$

$$= + \frac{\Gamma}{4\pi} \frac{z}{z^2 + (y - s)^2} \left[1 + \frac{-x}{\sqrt{(y - s)^2 + x^2 + z^2}} \right] \tag{11.13}$$

$$w_s = -q_s \cos \angle PEF$$

$$= - \frac{\Gamma}{4\pi} \frac{y - s}{z^2 + (y - s)^2} \left[1 + \frac{-x}{\sqrt{(y - s)^2 + x^2 + z^2}} \right] \tag{11.14}$$

And for the port vortex p we have

$$u_p = 0 \tag{11.15}$$

$$v_p = q_p \sin \angle PMF$$

$$= \frac{-\Gamma}{4\pi} \frac{z}{z^2 + (y + s)^2} \left[1 + \frac{-x}{\sqrt{(y + s)^2 + x^2 + z^2}} \right] \tag{11.16}$$

$$w_p = q_p \cos \angle PMF$$

$$= \frac{\Gamma}{4\pi} \frac{y + s}{z^2 + (y + s)^2} \left[1 + \frac{x}{\sqrt{(y + s)^2 + x^2 + z^2}} \right] \tag{11.17}$$

The total velocity components u, v, and w may now be found for points of interest about the wing through the use of Eqs. (11.1) to (11.3). However, the general solutions become involved, and it is preferable to limit our illustrations of the method to two zones:

1. The lateral plane ($x = 0$) and in particular the z axis
2. The longitudinal axis ($y = z = 0$)

These zones will be considered for both uniform and elliptic loading.

Many authors discuss induced velocities as a fraction of the induced velocity at the center of pressure of an elliptically loaded wing. The reason for this action is that dividing most of the expressions by $w_0 = (C_L/\pi \text{ A.R.}) \, V$ simplifies them considerably. However, the term

w_0 is much less familiar and easy to understand than the free-stream velocity V, and hence, although we shall write the equations in their simplest form, the illustrative examples and plots will assume an aspect ratio of 6.0 and $C_L = 1.0$, making it possible to give all induced velocities as a percentage of free-stream speed.

11.3. The Lateral Plane (Uniform Loading). The longitudinal component at the lateral plane is, from Eq. (11.9),

$$u = u_b = \frac{\Gamma}{4\pi z}\left[\frac{y + s}{\sqrt{(y + s)^2 + z^2}} - \frac{y - s}{\sqrt{(y - s)^2 + z^2}}\right] \quad (11.18)$$

or, for points on the z axis where $y = 0$, we have

$$u = \frac{\Gamma}{4\pi z}\frac{2s}{\sqrt{s^2 + z^2}} = \frac{SVC_L}{4s}\frac{1}{4\pi z}\frac{2s}{\sqrt{s^2 + z^2}}$$

$$= \frac{C_L}{\pi \text{ A.R.}}\frac{1}{2z}\frac{s^2}{\sqrt{z^2 + s^2}}V \quad (11.19)$$

and

$$\frac{u}{w_0} = \frac{1}{2z}\frac{s^2}{\sqrt{z^2 + s^2}} \quad (11.20)$$

As u is positive forward, the total velocity is $V - u$, and we see that V is increased above the wing and decreased below it. The actual

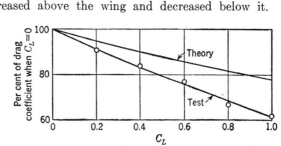

FIG. 11.3. Decrease of landing-gear drag coefficient with C_L for single-engine attack airplane. Part of the discrepancy is decrease in projected area with angle of attack; another part, failure of the vortex theory very close to the wing; and the remainder, difficulties of determining landing-gear drag center.

amount at a semi-span ($z = s$) below the wing is

$$\frac{u}{w_0} = \frac{1}{2s}\frac{s^2}{1.414s} = 0.353$$

For a wing of A.R. = 6 at the large value of $C_L = 1.0$,

$$w_0 = \frac{1.0}{6\pi}V \quad \text{and} \quad u = \frac{0.353}{6\pi}V = 0.0187V$$

or about 2 per cent of V. Thus from a first-approximation standpoint a pitot-static tube at the very large distance of one-half span below the wing on the z axis should reasonably yield about a 2 per cent error in free-stream velocity under the above uniform loading conditions. The decrease in wind velocity under the wing becomes much more pronounced as the wing is approached and, of course, as the lift coefficient increases. Thus, as is well known, the drag of the landing gear decreases more rapidly than the speed squared, and it is common to have the drag of the underwing mounting struts used for wind-tunnel models approach zero at high angles of attack.

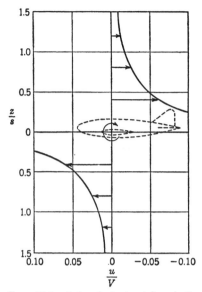

Fig. 11.4. Induced horizontal velocity along z axis, for A.R. = 6, C_L = 1.0, uniform loading.

Other values of u/V are shown in Fig. 11.4.

Continuing our discussion of the lateral plane ($x = 0$), we see that the lateral and vertical components are

$$v = v_p + v_s \tag{11.21}$$

and

$$w = w_p + w_s \tag{11.22}$$

Considering Eq. (11.21) first, we have, from Eqs. (11.13) and (11.16),

$$v = \frac{\Gamma}{4\pi} \frac{z}{z^2 + (y - s)^2} - \frac{\Gamma}{4\pi} \frac{z}{z^2 + (y + s)^2} \tag{11.23}$$

which clears to

$$v = \frac{\Gamma}{4\pi} \frac{4syz}{(y^2 + s^2 + z^2)^2 - 4s^2y^2} \tag{11.24}$$

Equation (11.22) becomes [after substitution from Eqs. (11.14) and (11.17)]

$$w = - \frac{\Gamma}{4\pi} \frac{(y - s)}{z^2 + (y - s)^2} + \frac{\Gamma}{4\pi} \frac{(y + s)}{z^2 + (y + s)^2} \tag{11.25}$$

and this clears to

$$w = - \frac{\Gamma}{4\pi} \frac{2s(y^2 - z^2 - s^2)}{(y^2 + z^2 + s^2)^2 - 4s^2y^2} \tag{11.26}$$

Substituting $\Gamma = SVC_L/4s$ and $w_0 = SVC_L/4\pi s^2$, we get

$$\frac{w}{w_0} = -\frac{1}{2}\frac{\dfrac{y^2}{s^2} - \dfrac{z^2}{s^2} - 1}{\left(\dfrac{y^2}{s^2} + \dfrac{z^2}{s^2} + 1\right)^2 - 4\dfrac{y^2}{s^2}} \tag{11.27}$$

Considering Eqs. (11.24) and (11.26), which represent the lateral and vertical components in the lateral ($x = 0$) plane, we see that, for nearly all values of y and z, the denominators are both positive. Hence the lateral velocity is directed toward the plane of symmetry above the wing

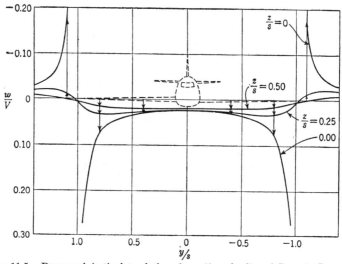

FIG. 11.5.   Downwash in the lateral plane for uniform loading, A.R. $= 6$, $C_L = 1.0$.

and away from the plane of symmetry below the wing. This is, of course, apparent from Fig. 11.1. The sign of the normal velocity w depends on the sign of the quantity $(y^2 - s^2 - z^2)$ and is in general such that w is directed downward below the wing and upward beyond the wing tips as in the Lanchester vortex pattern.

The upward flow beyond the wing tips is probably quite expected from a number of considerations: from the escape of the air at high pressure under the wing to the region of low pressure above the wing; or from the sense of the nearer tip vortex, which would have the preponderant effect. Of interest, too, is the somewhat similar situation that arises when a landing flap is lowered over the inboard portion of a wing. The drop in circulation at the most outboard flap station results in the shedding of a powerful trailing vortex having the same sense as the near tip vortex. The upward induced flow from this flap vortex increases the angle of attack of the wing outer panel, and in many cases of highly tapered wings a tip stall is thus produced that nullifies the beneficial effects of the flap.

Birds, incidentally, have realized the existence of upward flow beyond a wing tip and use it (in the Vee formation) when reduced effort is required.

Returning to our mathematical discussion, s becomes insignificant at great distances from the wing when compared with y and z, and Eqs. (11.24) and (11.26) then reduce to

$$v = \frac{\Gamma}{4\pi} \frac{4syz}{(y^2 + z^2)^2}$$

and

$$w = -\frac{\Gamma}{4\pi} \frac{2s(y^2 - z^2)}{(y^2 + z^2)^2} \tag{11.28}$$

Substituting $\Gamma = SVC_L/4s$, we get

$$v = \frac{SVC_L}{4s} \frac{1}{4\pi} \frac{4syz}{(y^2 + z^2)^2} = \frac{yz}{(y^2 + z^2)^2} \frac{SVC_L}{4\pi} \tag{11.29}$$

$$w = -\frac{SVC_L}{4s} \frac{1}{4\pi} \frac{2s(y^2 - z^2)}{(y^2 + z^2)^2} = \frac{-(y^2 - z^2)}{2(y^2 + z^2)^2} \frac{SVC_L}{4\pi} \tag{11.30}$$

and

$$\frac{w}{w_0} = -\frac{s^2}{2} \frac{(y^2 - z^2)}{(y^2 + z^2)^2} \tag{11.31}$$

Equations (11.29) and (11.30) will be of use later when wind-tunnel boundary effects are under scrutiny.

11.4. The Normal Axis (Uniform Loading). Reconsider the lateral-plane equations [Eqs. (11.26) and (11.24)] along the z axis (x and $y = 0$). We then have

$$v = 0$$

and

$$w = \frac{\Gamma}{4\pi} 2s \frac{(z^2 + s^2)}{(z^2 + s^2)^2} = \frac{\Gamma s}{2\pi(z^2 + s^2)} \tag{11.32}$$

Again introducing $\Gamma = SVC_L/4s$, we get

$$w = \frac{SVC_L}{4s} \cdot \frac{1}{2\pi} \cdot \frac{s}{z^2 + s^2} = \frac{1}{2} \frac{C_L}{\pi\,\text{A.R.}} \frac{s^2}{z^2 + s^2} \cdot V \tag{11.33}$$

Recalling that $w_0 = (C_L/\pi\,\text{A.R.})V$, we may rewrite Eq. (11.33) as

$$\frac{w}{w_0} = \frac{1}{2} \frac{1}{(z/s)^2 + 1} \tag{11.34}$$

Assuming several stations below the wing and an aspect ratio of 6 with $C_L = 1.0$, we find values shown in Table 11.1.

TABLE 11.1.   VALUES OF w/w_0 AND w/V ALONG THE z AXIS FOR UNIFORM AND ELLIPTIC
LOADING
(Elliptic loading values from Sect. 11.6.   Values of w/V are for A.R. = 6, $C_L = 1.0$.)

$\dfrac{z}{s}$	$\dfrac{w}{w_0}$ uniform loading	$\dfrac{w}{w_0}$ elliptic loading	$\dfrac{w}{V}$ uniform loading	$\dfrac{w}{V}$ elliptic loading
0	0.50	1.00	0.0265	0.053
0.25	0.471	0.758	0.0248	0.040
0.50	0.40	0.554	0.0212	0.0265
1.00	0.25	0.293	0.0133	0.0155
1.50	0.154	0.165	0.0082	0.0088
2.00	0.100	0.106	0.0053	0.0056
3.00	0.050	0.050	0.0027	0.00265
4.00	0.294	0.030	0.0016	0.0016

Of interest is the following: the downwash at the plane of symmetry of a uniformly loaded wing tends to 2.65 per cent V while the value for the elliptic wing is twice that, or 5.3 per cent V. At a semi-span below the wing the velocities on the z axis are quite similar. (This is not so for points out along the span, as will be seen in Sects. 11.3 and 11.7.)

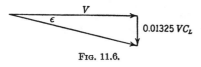

FIG. 11.6.

Another approach is to consider the local angle of flow deflection. For this case of the normal axis with uniform loading we have at a semi-span below the wing $w = 0.25w_0$ or $0.01325VC_L$. Thus the deflection angle* $\epsilon = 0.01325C_L$, or 0.01325 radian (= 45.5') at $C_L = 1.0$ (see Fig. 11.6). Downwash angles will assume greater importance in the following section.

11.5. The Longitudinal Axis (Uniform Loading). Now let us consider the longitudinal axis ($y = z = 0$). From the symmetry of the horseshoe vortex it is seen that along the x axis the lateral velocity $v = 0$.

---

* At the wing we use the symbol α_i for the downwash angle, while downstream from the wing the symbol ϵ is more common.

The normal component

$$w = w_b + w_s + w_p$$

$$= \frac{-\Gamma}{4\pi(-x)} \left(\frac{s}{\sqrt{x^2 + s^2}} - \frac{-s}{\sqrt{x^2 + s^2}} \right) + \frac{\Gamma}{4\pi s} \left(1 + \frac{-x}{\sqrt{x^2 + s^2}} \right)$$

$$+ \frac{\Gamma}{4\pi s} \left(1 + \frac{-x}{\sqrt{x^2 + s^2}} \right)$$

which clears to

$$w = \frac{\Gamma}{2\pi s} \left(1 - \frac{\sqrt{x^2 + s^2}}{x} \right)$$

Since x is positive forward, it is convenient to substitute $l = -x$ so that, finally,

$$\frac{w}{w_0} = \frac{1}{2} \left[1 + \sqrt{1 + \left(\frac{s}{l} \right)^2} \right] \tag{11.35}$$

Values of Eq. (11.35) with the usual A.R. $= 6$ and $C_L = 1.0$ are shown in Fig. 11.7.

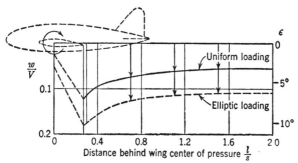

Fig. 11.7. Downwash velocity and angle behind wings of A.R. $= 6$ at $C_L = 1.0$. Elliptic-wing data from Sect. 11.9.

It is thus apparent that the flow field behind a uniformly loaded wing has a downward inclination which, since w_0 varies as C_L, varies with the lift. The local downwash angle is

$$\epsilon = \frac{w}{V} = \frac{C_L}{\pi \, \text{A.R.}} \cdot \frac{1}{2} \left[1 + \sqrt{1 + \left(\frac{s}{l} \right)^2} \right] \tag{11.36}$$

and the rate of change of downwash is

$$\frac{d\epsilon}{d\alpha} = \frac{1}{2} \frac{a}{\pi \, \text{A.R.}} \left[1 + \sqrt{1 + \left(\frac{s}{l} \right)^2} \right]$$

which is always less than 1.0—in practice about 0.4. Thus in effect the tail angle of attack will always change less than the wing angle, and in consequence the horizontal tail must be larger than preliminary estimates neglecting downwash would reveal.

Equation (11.36) also assumes paramount importance in considering the phenomenon of *tuck-under*, or *nose-heaviness*, which occurs at very high subsonic speeds. For reasons beyond the scope of this book, near-sonic phenomena can decrease the lift of the thick inboard stations of a wing without seriously affecting the outboard station. The loss of lift ahead of the tail results in a loss of downwash and a consequent increase

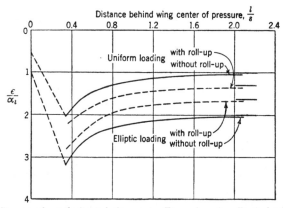

Fig. 11.8. Downwash angle behind the wing. The values for elliptic loading are from Sect. 11.9. The roll-up values for the uniform loading assume A.R. = 6.0.

of the tail angle of attack and tail lift. The added lift raises the tail and, unless provided for, can result in the destruction of the airplane through a subsequent dive.

Returning to Eq. (11.35), it would appear there that the downwash increases as the lifting line is approached until at the lifting line ($l = 0$) it is infinite. This effect should be recognized as the usual theoretical vortex failing. Figure 9.1 demonstrates that some boundary must be assigned for limiting the approach to the lifting line, and Sect. 11.4 indicates the proper values of downwash at $l = 0$. These limits have been applied in Figs. 11.7 and 11.8.

If we wish to compare the downwash with the induced angle at an elliptically loaded wing $\alpha_i = \dfrac{C_L}{\pi \text{ A.R.}}$, we have

$$\frac{\epsilon}{\alpha_i} = \frac{1}{2} [1 + \sqrt{1 + (s/l)^2}] \tag{11.37}$$

It will later be shown that an elliptically loaded wing should develop twice the downwash angle far aft of the center of pressure that it does *at* the center of pressure. Hence for that case ϵ/α_i should approach 2.0 aft of the wing. Equation (11.37) tends to approach 1.0 as $l \to \infty$.

11.6. The Vortex Span (Rectangular and Elliptic Wings). The induced velocity equations so far developed have been called *first-approximation values* since they contain no provision for the behavior of the trailing vortex sheet other than straight rearward travel. It is well known from both theory and practice that this does not occur. In the practical case there is always some decrease of circulation as the wing tip is approached, and hence many vortices are shed rather than the single one of ideal uniform loading. Thus there follows some roll-up of the shed vortices as discussed in Chap. 9, and, in addition, the whole vortex sheet is swept vertically downward by the bound vortex.

It develops that the roll-up takes place too slowly to affect seriously the amount of downwash in the vicinity of the horizontal tail, and the downward displacement of the vortex sheet moves the location of maximum induced velocities along with it without changing their values so that in the final analysis the induced velocities calculated without lateral or vertical displacement of the vortex sheet are quite in order as to magnitude, although somewhat in error as to location.

Investigators have found the actual vortex field to be as follows:

First of all a vortex appears immediately following the trailing edge of the wing, but its strength is somewhat less than that indicated by the maximum circulation at the wing mid-section. The reason for this is that some of the vortices, shed nearer to the plane of symmetry, have not yet had time to join the main tip vortex. This fact was demonstrated by emplacing a survey plane perpendicular to the tip vortex and selecting a closed path in that plane which encircled the vortex. The circulation found from $\oint \mathbf{q} \cdot d\mathbf{s}$ showed progressive increase as the survey plane moved farther and farther downstream. The reason, of course, was that additional vortices from farther inboard finally reached the tip vortex and added to its strength. The distance in which roll-up is supposed to be completed has been given (Ref. 11.6) as

$$x = 0.56s \frac{\text{A.R.}}{C_L} \tag{11.38}$$

where x = distance from quarter chord to a point where the roll-up is complete and the vortex span has become $2s'$, somewhat less than $2s$. For a lift coefficient of $C_L = 1$ and A.R. = 6 the roll-up would then take 3.36 semi-spans.

In experiments conducted by Silverstein, Katzoff, and Bullivant (Ref. 11.1) the roll-up of the vortices discussed in Sects. 9.2 and 11.9 was found to be of secondary importance as far as influencing the flow at a reasonable tail plane position is concerned as long as the aspect ratio is

228   *BASIC WING AND AIRFOIL THEORY*

large.* But the downward displacement of the vortex sheet due to the bound vortex had to be considered. This displacement may be readily calculated, since the centerline of the vortex sheet passes through the wing trailing edge and moves downward with the downwash itself. Its inclination at every point is the downwash angle ϵ at that point, and hence the vertical displacement h at some point aft of the trailing edge is

$$h = \int_{T.E.}^{x} \tan \epsilon \, dx \qquad (11.39)$$

and the trailing vortex is hence slightly curved.

Fortunately, it is not necessary to consider the curvature, and adequately accurate downwash calculations accrue from considering the vortex as not rolled up, but displaced downward. In other words, the downwash velocities and angles along the longitudinal axis as calculated by the method outlined in this chapter match accurately those found in practice along a longitudinal axis that has been displaced in accordance with Eq. (11.39).

Returning to the problem of the downwash velocities and angles along the longitudinal axis, we may investigate the roll-up by assuming that the final distance between the vortices is $2s'$. Since only two vortices then exist, it is in order to assume that they are due to a uniform circulation Γ_a over a wing of the shortened span $2s'$. Then

$$L = \rho V \Gamma_a (2s')$$

and expressing the actual distribution of circulation as

$$\Gamma = 4sV \sum_{1}^{\infty} A_n \sin n\theta$$

we have from Eq. (10.13)

$$L = 2\pi s^2 \rho V^2 A_1$$

Along the longitudinal axis $\theta = 90°$, and hence the circulation series becomes

$$\Gamma = 4sV(A_1 - A_3 + A_5 - \cdots)$$

$$\frac{s'}{s} = \frac{L}{2\rho V \Gamma_a} \cdot \frac{2\pi s \rho V^2 A_1}{L}$$

$$= \frac{\pi}{4} \frac{A_1}{A_1 - A_3 + A_5} \qquad (11.40)$$

For elliptic loading where only A_1 exists, we have immediately that the final vortex span is 78.5 per cent of the actual span regardless of aspect ratio. From the values of $A_1/\mu\alpha$ in Fig. 10.5 we may calculate

* Recently wings of small aspect ratio have come into prominence. For this case the work of Spreiter and Sacks (Ref. 11.8) should be consulted.

the vortex span for rectangular wings, which, obviously, varies with aspect ratio and is generally greater than that for elliptic loading. Values of s'/s are shown in Fig. 11.9 for elliptic and tapered wings. They should be considered instead of the geometric span for problems where

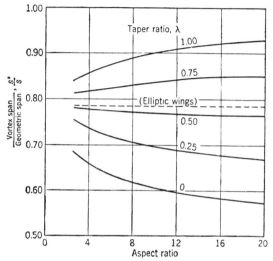

Fɪɢ. 11.9. Values of the vortex span for elliptic, rectangular, and tapered wings.

the downstream effect of the vortices is being found, an example being the calculation of wind-tunnel wall corrections.*

The effect of roll-up on the downwash follows: From Eq. (11.35),

$$w = \frac{\Gamma_a}{2\pi s'}\left[1 + \frac{\sqrt{l^2 + (s')^2}}{l}\right]$$
$$= \frac{L}{4\pi(s')^2\rho V}\left[1 + \frac{\sqrt{l^2 + (s')^2}}{l}\right]$$

and the downwash becomes

$$\epsilon = \frac{w}{V} = \frac{(\rho/2)SV^2C_L}{4\pi(s')^2\rho V^2}\left[1 + \frac{\sqrt{l^2 + (s')^2}}{l}\right]$$
$$= \frac{C_L S}{2\pi \cdot 4s^2}\left(\frac{s}{s'}\right)^2\left[1 + \frac{\sqrt{l^2 + (s')^2}}{l}\right]$$
$$\frac{\epsilon}{\alpha_i} = \frac{1}{2}\left(\frac{s}{s'}\right)^2\left[1 + \frac{\sqrt{l^2 + (s')^2}}{l}\right] \qquad (11.41)$$

* For wind-tunnel wall corrections concerning the main wing, the vortex span far downstream does not have the importance that it has, say, near the tail, and a reasonable compromise is to use 0.9 span instead of theoretical values closer to 0.8.

230   BASIC WING AND AIRFOIL THEORY

As $l \to \infty$, the limiting value of the downwash becomes

$$\frac{\epsilon}{\alpha_i} = \frac{1}{2}\left(\frac{s}{s'}\right)^2 \left[1 + \sqrt{1 + \left(\frac{s'}{l}\right)^2}\right] = \left(\frac{s}{s'}\right)^2 \qquad (11.42)$$

Frequently, however, the airplane designer seeks the rate of change of downwash with angle of attack $d\epsilon/d\alpha$ rather than the vortex span. Thus

$$\epsilon = \frac{1}{2}\frac{C_L}{\pi \text{ A.R.}}\left(\frac{s}{s'}\right)^2\left[1 + \sqrt{\frac{l^2 + (s')^2}{l^2}}\right]$$

$$\frac{d\epsilon}{d\alpha} = \frac{1}{2}\left(\frac{s}{s'}\right)^2\left[1 + \sqrt{\frac{l^2 + (s')^2}{l^2}}\right]\frac{dC_L}{d\alpha}\frac{1}{\pi \text{ A.R.}}$$

Or, from Eq. (11.42),

$$\frac{d\epsilon}{d\alpha} = \frac{\epsilon}{\alpha_i}\frac{a}{\pi \text{ A.R.}}$$

Values of $d\epsilon/d\alpha$ are shown in Fig. 11.10.

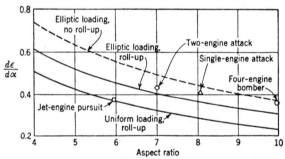

Fig. 11.10.   Values of $d\epsilon/d\alpha$, theory and flight test.   The theoretical values are for the x axis one semi-span behind the wing.   (Elliptic-loading figures are from Sect. 11.9.)   The flight test points are for various tail lengths and merely demonstrate reasonable values.

We shall now examine induced flow along the various axes for the case of elliptic loading.

11.7. The Normal Axis (Elliptic Loading).   For the case of elliptic loading the circulation at any point on the span is, from Eq. (9.33),

$$\Gamma = \Gamma_0 \sqrt{1 - \left(\frac{2y}{b}\right)^2}$$

where Γ_0 = value of Γ at mid-span.
Using Γ_0 from Eq. (9.41) and L from Eq. (10.13), we have, letting $s = b/2$,

$$\Gamma = 2VbA_1 \sqrt{1 - \left(\frac{y}{s}\right)^2}$$

Since, for elliptic loading,

$$w_0 = \frac{\Gamma_0}{2b} = \frac{2bVA_1}{2b} = VA_1$$

then

$$\Gamma = 4w_0 \sqrt{s^2 - y^2}$$

and

$$\frac{\partial \Gamma}{\partial y} = \frac{4w_0 y}{\sqrt{s^2 - y^2}}$$

Now the expressions for the induced velocity at various locations have already been determined for the case of uniform loading, the integration in each case having been performed all the way across the span so that the complete effects are contained in each expression. We will now simulate elliptic loading by summing a number of uniform loadings as shown in Fig. 9.4. Each will have a strength of $\frac{\partial \Gamma}{\partial y} dy$ and we will only need to integrate from 0 to s in order to determine the total number, or, in other words, Γ. Thus the expression for the induced velocity along the z axis has already been determined for the case of uniform loading (Eq. 11.32) as

$$w = \frac{\Gamma}{2\pi} \frac{s}{z^2 + s^2}$$

for uniform loading. Hence for elliptic loading

$$w = \int_0^s \frac{1}{2\pi} \left(-\frac{\partial \Gamma}{\partial y} dy \right) \frac{y}{z^2 + y^2}$$

or

$$w = \frac{1}{2\pi} \int_0^s \frac{4w_0 y}{\sqrt{s^2 - y^2}} \frac{y}{z^2 + y^2} dy \tag{11.43}$$

Letting $y = s \sin \theta$, $dy = s \cos \theta\, d\theta$, we have

$$\frac{w_0}{w} = \frac{2}{\pi} \int_0^{\pi/2} \frac{s^2 \sin^2 \theta\, d\theta}{z^2 + s^2 \sin^2 \theta}$$

Dividing numerator by denominator

$$\frac{w}{w_0} = \frac{2}{\pi} \int_0^{\pi/2} \left(1 - \frac{z^2}{z^2 + s^2 \sin^2 \theta} \right) d\theta$$

$$= \frac{2}{\pi} \cdot \frac{\pi}{2} - \frac{2}{\pi} \int_0^{\pi/2} \frac{(z/s)^2}{(z/s)^2 + \sin^2 \theta} d\theta$$

Considering only the integral, we have

$$\frac{2}{\pi} \frac{z^2}{s^2} \int_0^{\pi/2} \frac{2d\theta}{2(z^2/s^2) + 1 - \cos 2\theta} \tag{11.44}$$

Equation (11.44) is directly integrable by Formula 300 of Ref. 11.2, which is

$$\int \frac{dx}{a + b \cos x} = \frac{2}{\sqrt{a^2 - b^2}} \tan^{-1} \frac{\sqrt{a^2 - b^2} \tan (x/2)}{(a + b)}$$

Letting $a = (2z^2/s^2) + 1$, $b = -1$, $x = 2\theta$, $dx = 2d\theta$, we have

$$\frac{2}{\pi} \frac{z^2}{s^2} \int_0^{\pi/2} \frac{2d\theta}{2\left(\dfrac{z}{s}\right)^2 + 1 - \cos 2\theta} = \frac{2}{\pi} \frac{z^2}{s^2} \left[\frac{2}{\sqrt{\left(2\dfrac{z^2}{s^2} + 1\right)^2 - (-1)^2}} \right.$$

$$\left. \cdot \tan^{-1} \frac{\sqrt{\left(2\dfrac{z^2}{s^2} + 1\right)^2 - 1} \cdot \tan \theta}{\left(2\dfrac{z^2}{s^2} + 1\right) + (-1)} \right]_0^{\pi/2}$$

Substituting the limits, we have $\tan 0 = 0$ and $\tan^{-1} 0 = 0$ and, at $\pi/2$, $\tan (\pi/2) = \infty$ and $\tan^{-1} \infty = \pi/2$. Hence

$$\frac{2}{\pi} \frac{z^2}{s^2} \int_0^{\pi/2} \frac{2d\theta}{2\left(\dfrac{z}{s}\right)^2 + 1 - \cos 2\theta} = \frac{2^2}{\pi^2} \frac{z^2}{s^2} \frac{2}{\sqrt{\left(2\dfrac{z^2}{s^2} + 1\right)^2 - 1}} \cdot \frac{\pi}{2}$$

and

$$\frac{w}{w_0} = 1 - \frac{2z^2}{s^2} \frac{1}{\sqrt{4(z^4/s^4) + 4(z^2/s^2) + 1} - 1}$$

$$= 1 - \frac{z/s}{\sqrt{(z/s)^2 + 1}} \qquad (11.45)$$

Values for Eq. (11.45) are given in Table 11.1. As a matter of interest, the difference in downwash angle between elliptic and uniform loading at a semi-span below the wing is but $0.127°$ for a wing of A.R. = 6 at $C_L = 1.0$.

Rewriting Eq. (11.45) as

$$\frac{w}{w_0} = \frac{\sqrt{z^2 + s^2} - z}{\sqrt{z^2 + s^2}}$$

$$\left(\frac{w}{w_0}\right)^2 = \frac{z^2 + s^2 - 2z\sqrt{z^2 + s^2} + z^2}{z^2 + s^2}$$

Now, expanding the radical as a binomial,

$$\left(\frac{w}{w_0}\right)^2 = \frac{2z^2 + s^2 - 2z\left(z + \dfrac{s^2}{2z} - \dfrac{s^4}{8z^3} \cdots\right)}{z^2 + s^2}$$

$$= \frac{s^4}{4z^4 + 4z^2s^2}$$

For large z, s becomes small against z, and we may drop $4z^2s^2$ against $4z^4$. Then

$$\frac{w}{w_0} = \frac{s^2}{2z^2}$$

Since

$$w_0 = \frac{C_L}{\pi \text{ A.R.}} \, V = \frac{SVC_L}{4\pi s^2}$$

we have

$$w = \frac{SVC_L}{8\pi z^2}$$

Referring to Sect. 11.4 and neglecting s^2 against z^2 as above, we find that Eq. (11.34) reverts to the equation above, demonstrating that at large

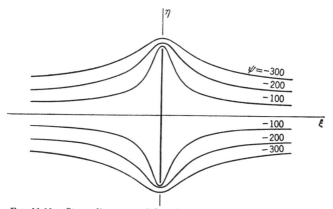

FIG. 11.11.   Streamlines around flat plate at right angles to free stream.

distances from the wing the induced velocity is independent of the span loading.

11.8. The Lateral Plane (Elliptic Loading).   The principle already developed that elliptic loading yields uniform downwash w_0 across the span may now be utilized to yield the flow pattern in the lateral ($x = 0$) plane.

To review, the streamlines past a flat plate at 90° to a uniform flow are shown in Fig. 11.11.   The flow is expressed by the potential function

$$w = \phi + i\psi = iV \sqrt{\zeta^2 - 4a^2} \tag{11.46}$$

where $\zeta = \xi + i\eta$ and the width of the line is $4a$.   The stream function of the flow is ψ, and the vertical velocity (positive up) at any point in the field is $-\dfrac{\partial \psi}{\partial \xi}$.

Figure 11.11 shows the flow as if the plate were still and the fluid moving. Consider the case, starting from rest, where the fluid is still and the plate moves. Particles (·) ahead of the fluid are pushed along the stream while particles behind the plate (∘) are carried along. A certain tip flow develops, and the particle paths (the streamlines) appear as in Fig. 11.12. The student may be reminded of the view downward from the front of a scow.

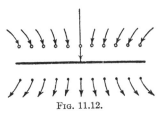

FIG. 11.12.

Now in the case of the airplane wing with elliptic loading the airplane wing produces a uniform downwash along the span. Relative to the fluid the figure resembles Fig. 11.12, and at the wing the downward velocity $w = w_0$.

To return to the expression for this flow, we already have the local vertical velocity for Fig. 11.11,

$$w = -\frac{\partial \psi}{\partial \xi}$$

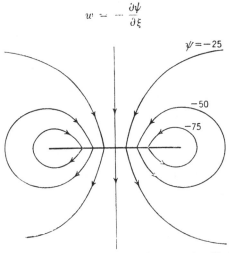

FIG. 11.13. Streamlines of flat plate moving in still air.

Recalling that this is relative to the flat plate, we may find it relative to the fluid by simply adding the vertical velocity w_0, having finally

$$w = w_0 - \frac{\partial \psi}{\partial \xi} \tag{11.47}$$

This mechanism is further explained in Fig. 11.13. Numerical values later demonstrated show that $-\dfrac{\partial \psi}{\partial \xi}$ yields an upward flow as in Fig. 11.11.

Before we start the analysis of the problem, three changes in Eq. (11.46) must be made: First, since we are currently using the symbol w for downwash, we shall use W for the potential function. Second, the uniform vertical velocity V becomes the uniform downwash w_0. Third, the semi-span $s = 2a$. Finally

$$W = iw_0 \sqrt{\zeta^2 - s^2} \qquad (11.48)$$

Equation (11.48) is not yet useful, however, as in order to obtain ϕ and ψ we must separate the potential function into reals and imaginaries. This, it develops, is most simply done by making the substitution

$$\zeta = s \sin (\lambda + i\mu)$$
$$= s \sin \lambda \cos i\mu + s \cos \lambda \sin i\mu$$

Using Eqs. (2.33), we have

$$\zeta = s \sin \lambda \cosh \mu + is \cos \lambda \sinh \mu$$

or

$$\xi = s \sin \lambda \cosh \mu$$
$$\eta = s \cos \lambda \sinh \mu \qquad (11.49)$$

Equation (11.48) becomes

$$W = iw_0 \sqrt{\zeta^2 - s^2}$$
$$= iw_0 s \sqrt{\sin^2 (\lambda + i\mu) - 1}$$
$$= -w_0 s \cos (\lambda + i\mu) \qquad (11.50)$$

Separating the potential function into stream function and velocity potential, we have

$$W = \phi + i\psi$$
$$= -w_0 s (\cos \lambda \cos i\mu - \sin \lambda \sin i\mu)$$
$$= w_0 s (\cos \lambda \cosh \mu - i \sin \lambda \sinh \mu)$$

so that

$$\phi = -w_0 s \cos \lambda \cosh \mu$$
$$\psi = w_0 s \sin \lambda \sinh \mu$$

Equation (11.47) becomes

$$w = w_0 - \frac{\partial \psi}{\partial \xi}$$
$$= w_0 - \frac{\partial (w_0 s \sin \lambda \sinh \mu)}{\partial (s \sin \lambda \cosh \mu)} \qquad (11.51)$$

From fundamental laws of partials

$$\frac{\partial \psi}{\partial \xi} = \frac{\partial \psi}{\partial \lambda} \frac{\partial \lambda}{\partial \xi} + \frac{\partial \psi}{\partial \mu} \frac{\partial \mu}{\partial \xi} \qquad (11.52)$$

Using Eqs. (11.49),

$$\frac{\partial \xi}{\partial \xi} = 1 = s \cos \lambda \cosh \mu \frac{\partial \lambda}{\partial \xi} + s \sin \lambda \sinh \mu \frac{\partial \mu}{\partial \xi}$$

$$\frac{\partial \eta}{\partial \xi} = 0 = -s \sin \lambda \sinh \mu \frac{\partial \lambda}{\partial \xi} + s \cos \lambda \cosh \mu \frac{\partial \mu}{\partial \xi}$$

from which

$$\frac{\partial \mu}{\partial \xi} = \frac{\sin \lambda \sinh \mu}{\cos \lambda \cosh \mu} \frac{\partial \lambda}{\partial \xi} \tag{11.53}$$

and

$$\frac{\partial \xi}{\partial \xi} = 1 = s \cos \lambda \cosh \mu \frac{\partial \lambda}{\partial \xi} + s \sin \lambda \sinh \mu \cdot \frac{\sin \lambda \sinh \mu}{\cos \lambda \cosh \mu} \frac{\partial \lambda}{\partial \xi}$$

or

$$\frac{\partial \lambda}{\partial \xi} = \frac{\cos \lambda \cosh \mu}{s(\cos^2 \lambda \cosh^2 \mu + \sin^2 \lambda \sinh^2 \mu)}$$

so that Eq. (11.53) becomes

$$\frac{\partial \mu}{\partial \xi} = \frac{\sin \lambda \sinh \mu}{s(\cos^2 \lambda \cosh^2 \mu + \sin^2 \lambda \sinh^2 \mu)}$$

Also,

$$\frac{\partial \psi}{\partial \lambda} = s \cos \lambda \sinh \mu \cdot w_0$$

$$\frac{\partial \psi}{\partial \mu} = s \sin \lambda \cosh \mu \cdot w_0$$

Substituting in Eq. (11.52), we have

$$\frac{\partial \psi}{\partial \xi} = \frac{w_0 s \cos \lambda \sinh \mu \cos \lambda \cosh \mu}{s(\cos^2 \lambda \cosh^2 \mu + \sin^2 \lambda \sinh^2 \mu)} + \frac{w_0 s \sin \lambda \cosh \mu \sin \lambda \sinh \mu}{s(\cos^2 \lambda \cosh^2 \mu + \sin^2 \lambda \sinh^2 \mu)}$$

Combining terms, and using $\cos^2 \lambda + \sin^2 \lambda = 1$, we have

$$\frac{\partial \psi}{\partial \xi} = w_0 \frac{\sinh \mu \cosh \mu}{\cos^2 \lambda \cosh^2 \mu + \sin^2 \lambda \sinh^2 \mu}$$

And substituting $\cos^2 \lambda = 1 - \sin^2 \lambda$ and $\sinh^2 \mu = \cosh^2 \mu - 1$, we have

$$\frac{\partial \psi}{\partial \xi} = \frac{\sinh \mu \cosh \mu}{\cosh^2 \mu - \sin^2 \lambda}$$

And finally Eq. (11.47) is

$$w = w_0 \left(1 - \frac{\sinh \mu \cosh \mu}{\cosh^2 \mu - \sin^2 \lambda}\right) \tag{11.54}$$

Now the symbols ξ and η of Eq. (11.49) correspond, respectively, to y and z, and hence we may substitute selected values of y/s and z/s into Eq. (11.49) and, solving Eq. (11.54), get values for w/w_0.

For example, when $y/s = 0.25$ and $z/s = 0.10$, we get from Eq. (11.49)

$$\sin \lambda = \frac{0.25}{\cosh \mu}$$

$$\cos \lambda = \frac{0.10}{\sinh \mu}$$

Solving,

$$\sinh \mu = 0.103$$
$$\cosh \mu = 1.005$$

and

$$\sin \lambda = 0.248$$

Substituting in Eq. (11.54),

$$\frac{w}{w_0} = 1 - 0.111 = 0.889$$

Further values are shown in Fig. 11.14.

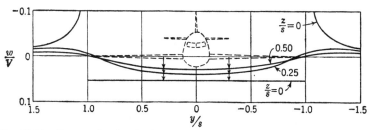

FIG. 11.14.   Downwash velocities in yz plane for elliptic loading, A.R. $= 6$, $C_L = 1.0$.

11.9. The Longitudinal Axis (Elliptic Loading).   The expression for the downwash on the longitudinal axis for uniform loading [Eq. (11.35)] may be extended to cover elliptic loading by substitution of the proper variables.   Hence

$$\epsilon = \frac{\Gamma}{2\pi s V} \left(1 + \frac{\sqrt{l^2 + s^2}}{l} \right) \tag{11.55}$$

$$= \int_0^s \frac{1}{2\pi y V} \cdot \frac{4 w_0 y}{\sqrt{s^2 - y^2}} \left(1 + \frac{\sqrt{l^2 + y^2}}{l} \right) dy$$

Writing $\alpha_i = w_0/V$, we have

$$\epsilon = \frac{2}{\pi} \alpha_i \int_0^s \left(1 + \frac{\sqrt{l^2 + y^2}}{l} \right) \frac{y}{s} \frac{dy}{\sqrt{s^2 - y^2}} \tag{11.56}$$

Letting $y = s \cos \theta$, $dy = -s \sin \theta \, d\theta$ and changing limits,

$$\frac{\epsilon}{\alpha_i} = \frac{2}{\pi} \int_0^{\pi/2} \left(1 + \frac{\sqrt{l^2 + s^2 \cos \theta}}{l} \right) d\theta \tag{11.57}$$

To simplify the integration, write a distance downstream in terms of the semi-span according to $l^2 = \dfrac{1 - k^2}{k^2} s^2$, where k is a variable. Then

$$\frac{\epsilon}{\alpha_i} = 1 + \frac{2}{\pi} \frac{E}{\sqrt{1 - k^2}} \tag{11.58}$$

where E is the complete elliptic integral

$$E = \int_0^{\pi/2} \sqrt{1 - k^2 \sin^2 \theta} \, d\theta$$

Selecting values of l/s, we may find k and then E from the tables of Ref. 11.2. The results are plotted in Fig. 11.8.

Using the second-approximation logic of Sect. 11.5, Eq. (11.40) reduces (for the case of elliptic loading) to

$$\frac{s'}{s} = \frac{\pi}{4} = 0.785$$

and the value of ϵ/α_i far behind the wing becomes, from Eq. (11.41),

$$\left(\frac{\epsilon}{\alpha_i}\right)_{\substack{\lim \\ l \to \infty}} = \left(\frac{s}{s'}\right)^2 = \frac{1}{0.785^2} = 1.63$$

as compared with 2.0 for Eq. (11.57). As before,

$$\frac{\epsilon}{\alpha_i} = \frac{1}{2} \left(\frac{s}{s'}\right)^2 \left[1 + \frac{\sqrt{l^2 + (s')^2}}{l}\right] \tag{11.59}$$

At $l = s$,

$$\begin{aligned} \frac{\epsilon}{\alpha_i} &= \frac{1}{2} \left(\frac{s}{s'}\right)^2 \left[1 + \sqrt{1 + \frac{(s')^2}{s^2}}\right] \\ &= \tfrac{1}{2}(1.63)(1 + 1.27) \\ &= 1.85 \end{aligned}$$

and

$$\begin{aligned} \frac{d\epsilon}{d\alpha} &= \frac{\epsilon}{\alpha_i} \frac{a}{\pi \, (\text{A.R.})} \\ &= 1.85 \cdot \frac{4.71}{6\pi} = 0.46 \end{aligned}$$

for A.R. = 6.0.

Values of Eq. (11.59) are shown in Fig. 11.8, and one is struck by the reduction in downwash on the x axis according to whether the roll-up is considered to occur or not. From a cursory examination it would appear that the downwash should remain the same. Actually the total downward energy does remain the same, but the spanwise distribution of downwash for the rolled-up condition is that of uniform loading (Fig. 11.5), while that for the unrolled-up condition is that of elliptic loading (Fig. 11.14). The final result for the rolled-up condition yields less downwash at the x axis but more further outboard.

We may now compare theory and practice for the downwash behind a wing in order to ascertain the degree of accuracy of the theory. Silverstein in Ref. 11.1 has measured values for a Clark Y wing of aspect ratio 6 at a $C_L = 0.74$. We shall use his data for comparison. First of all, to obtain the theoretical values, we shall need ϵ/α_i for the elliptic loading case (no roll-up) and for the uniform loading case (no roll-up). These we get from Fig. 11.8. Since $\alpha_i = C_L/\pi$ A.R. $= (0.74/6\pi)57.3 = 2.25°$, we finally get the downwash angles for the two loadings, as in Table 11.2.

TABLE 11.2

$\dfrac{l}{s}$	$\left(\dfrac{\epsilon}{\alpha_i}\right)_{\text{elliptic}}$	$\epsilon_{\text{elliptic}}$	$\left(\dfrac{\epsilon}{\alpha_i}\right)_{\text{uniform}}$	$\epsilon_{\text{uniform}}$	ϵ_{mean}
0.4	2.95	6.63	1.85	4.16	5.40
0.8	2.33	5.24	1.30	2.93	4.09
1.2	2.17	4.88	1.16	2.61	3.74
1.6	2.11	4.74	1.08	2.43	3.58

According to Schrenk (Ref. 12.3) a good approximation to the spanwise lift distribution is to take a mean between the distribution of chord

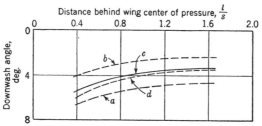

Fig. 11.15. Comparison of no roll-up downwash values with experiment. a, downwash angle for elliptic loading, no roll-up [Eq. (11.58)]; b, downwash angle for uniform loading, no roll-up [Eq. (11.36)]; c, mean of $a + b$; d, values from experiment.

and an equivalent ellipse (see Chap. 12), and correspondingly the downwash for a rectangular wing becomes a mean between the downwash for uniform loading and that for elliptic loading. This mean curve is shown on Fig. 11.15. Comparing the experimental curve with the theory, we find agreement within $\frac{1}{2}°$ in the tail region—certainly a satisfactory check.

11.10. Sweepback (Uniform Loading). Sometimes, in order to achieve balance (by moving the mean aerodynamic chord without moving the root chord) or to reduce the effects of compressibility at high speeds, a wing will have sweepback. Its lifting line, then, will have sweepback too, and the simple vortex pattern previously discussed will not be realized.

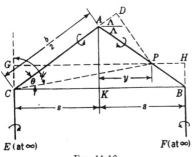

Fig. 11.16.

To understand the effect of sweepback, consider the bent lifting line with uniform circulation shown in Fig. 11.16. Since a straight vortex segment has no effect on itself, the downwash velocity at P will be due to the bound lifting-line segment AC and the port and starboard trailing vortices. Thus

$$w = w_b + w_p + w_s \tag{11.60}$$

Considering w_b first, we have from the Biot-Savart law

$$w_b = \frac{\Gamma}{4\pi PD} \tag{11.61}$$

Now

$$PD = y \sec \Lambda (\sin 2\Lambda) = 2y \sin \Lambda$$

and

$$AD = y \sec \Lambda (\cos 2\Lambda) = y(\cos \Lambda - \sin \Lambda) \tag{11.62}$$

Letting $\angle PCA$ be called θ, we have

$$\tan PCA = \tan \theta = \frac{PD}{AC + AD} = \frac{2y \sin \Lambda}{s \sec \Lambda + y(\cos \Lambda - \sec \Lambda)}$$

which clears to

$$\tan \theta = \frac{y/s \sin 2\Lambda}{1 + y/s \sin^2 \Lambda} \tag{11.63}$$

From the figure

$$\cos \angle PAC = -\cos 2\Lambda$$

Considering one-half of the wing, we may write its lift as

$$\frac{L}{2} = \frac{\rho}{2} SV^2 \frac{C_L}{2} = \rho (V \cos \Lambda) \frac{b}{2} \Gamma$$

which, since $b/2 = s \sec \Lambda$, may be reduced to

$$\Gamma = \frac{SVC_L}{2} \tag{11.64}$$

The aspect ratio is $4s^2/S$, so that finally

$$\Gamma = \frac{sVC_L}{\text{A.R.}} \tag{11.65}$$

and Eq. (11.61) becomes

$$w_b = \frac{VC_L \csc \Lambda}{8\pi(y/s) \text{ A.R.}} (\cos \theta - \cos 2\Lambda)$$

Now consider the downwash at P due to the port trailing vortex CE

$$w_p = \frac{\Gamma}{4\pi PG} (\cos \angle PEC - \cos \angle PCG) \tag{11.66}$$

We have

$$PG = y + s$$

and letting

$$\angle PCG = \angle \psi$$

$$\tan \psi = \frac{s + y}{(s - y) \tan \Lambda}$$

so that

$$\psi = \tan^{-1} \left[\frac{1 + y/s}{1 - y/s} \cot \Lambda \right] \tag{11.67}$$

and finally

$$w_p = \frac{VC_L}{4\pi \text{ A.R.}} \frac{1 - \cos \psi}{1 + y/s} \tag{11.68}$$

The downwash at P due to the starboard vortex is

$$w_s = \frac{\Gamma}{4\pi PH} (\cos \angle PFB - \cos \angle PBH) \tag{11.69}$$

$$w_s = \frac{\Gamma}{4\pi(s - y)} (1 - \sin \Lambda)$$

$$w_s = \frac{VC_L}{4\pi \text{ A.R.}} \frac{(1 - \sin \Lambda)}{(1 - y/s)} \tag{11.70}$$

The total downwash is then

$$\frac{w_T}{V} = \frac{C_L}{4\pi \text{ A.R.}} \left[\frac{\csc \Lambda}{2y/s} (\cos \theta - \cos 2\Lambda) + \frac{1 - \cos \psi}{1 + y/s} + \frac{1 - \sin \Lambda}{1 - y/s} \right] \tag{11.71}$$

The following example illustrates how sweepback reduces the angle of attack near the centerline and thus moves the centroid of lift out along the span:

Example 11.1. Find the distribution of downwash along a uniformly loaded swept-back wing if A.R. = 6.0, C_L = 1.0, and Λ = 40°.

1. $\dfrac{C_L}{4\pi \text{ A.R.}} = \dfrac{1.0}{4\pi \cdot 6} = 0.01323$

2. $\Lambda = 40°$, $\sin \Lambda = 0.643$, $\cos \Lambda = 0.766$, $\csc \Lambda = 1.556$, $\cot \Lambda = 1.192$, $2\Lambda = 80°$, $\sin 2\Lambda = 0.985$, $\cos 2\Lambda = 0.174$

3. $\theta = \tan^{-1} \dfrac{0.985y/s}{1 + 0.413y/s}$; $\psi = \tan^{-1} 1.192 \left(\dfrac{1 + y/s}{1 - y/s}\right)$

4. $w_b = 0.01323 \dfrac{1.556}{2y/s} (\cos \theta - \cos 2\Lambda)$

   $= \dfrac{0.0103(\cos \theta - 0.174)}{y/s}$

5. $w_p = 0.01323 \dfrac{(1 - \cos \psi)}{1 + y/s}$

6. $w_s = 0.01323 \dfrac{1 - 0.643}{1 - y/s} = \dfrac{0.00473}{1 - y/s}$

7. The complete results are shown in Fig. 11.17 and Table 11.3.

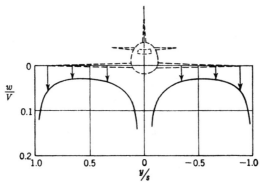

Fig. 11.17. Downwash along the bent lifting line of a uniformly loaded swept-back wing. $C_L = 1.0$, A.R. $= 6$, $\Lambda = 40°$.

TABLE 11.3

$\dfrac{y}{s}$	θ	ψ	$\dfrac{w_b}{V}$	$\dfrac{w_p}{V}$	$\dfrac{w_s}{V}$	$\dfrac{w_T}{V}$
0	0	50°	∞	0.0047	0.0047	∞
0.1	5°20′	50°42′	0.0843	0.0045	0.0053	0.0940
0.2	10°20′	56°18′	0.0418	0.0049	0.0059	0.0526
0.4	18°40′	66°48′	0.0199	0.0058	0.0079	0.0335
0.6	25°21′	75°58′	0.0126	0.0063	0.0118	0.0306
0.8	30°39′	83°40′	0.0088	0.0066	0 0236	0.0390
0.9	32°39′	87°0	0.0077	0.0066	0 0473	0.0616
1.0	34°50′	90°	0.0067	0.0069	∞	∞

11.11. Dihedral (Uniform Loading). Early experimenters found that putting a dihedral angle in the wing greatly improved lateral stability through providing a tendency for the airplane to right itself after a side-slip. The actual amount of roll developed is, in the practical case, a function of many factors: the dihedral, and where it starts on the wing; whether the airplane is a high-wing or low-wing design; and the wing taper and sweep. However, the mechanism of the simple case of the straight wing with dihedral makes an interesting starting point.

From Fig. 11.18 it is seen that a straight wing with dihedral develops a greater angle of attack on the windward side in a slip and a lesser amount on the leeward side. The change in angle is quite difficult to demon-

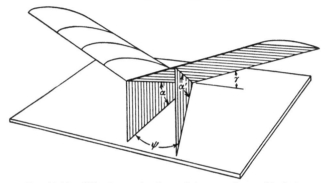

Fig. 11.18.   Effective angle-of-attack increase due to dihedral.

strate without a model, and suffice it here merely to state that a rectangular wing with a dihedral angle γ, a yaw angle ψ, and a root angle of attack α will develop an angle of attack on port and starboard wings in the amount

$$\alpha' = \sec^{-1} \sqrt{1 + \sin^2 \psi (\tan \gamma + \cot \psi \tan \alpha)^2} \qquad (11.72)$$

If we suppose the wing to be uniformly loaded (an approximation, of course) then the lift increment ΔL will act at the mid-point of the semi-span and roll moment developed will be (for two half wings)

$$RM = \Delta L \frac{s}{2} \cdot 2$$

From the definition $C_l = RM/qSb$, $\Delta \alpha_d = \alpha' - \alpha$, and $\Delta C_L = \Delta \alpha_d \cdot a$ (where a is the slope of the lift curve), we get

$$C_l = \frac{\Delta \alpha_d \cdot a}{2qS} \cdot q \cdot \frac{S}{2} = a \frac{\Delta \alpha_d}{4} \qquad (11.73)$$

We may then use Eq. (10.35) to determine the lift curve slope for a particular aspect ratio and find the roll due to dihedral thereby. This process has been followed for the case of A.R. = 6 and the results pre-

sented in Fig. 11.19.   To a close approximation this elementary method
yields a roll of $dC_l/d\psi$ = 0.00027 per degree of dihedral.   In practice,

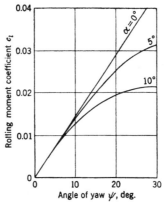

Fig. 11.19.   Effect of 5° dihedral on rolling-moment coefficient due to yaw.   A.R. = 6.0,
uniform loading.

less than the above value has been found, and it is customary to call
$dC_l/d\psi$ = 0.00021 "one degree of effective dihedral."

PROBLEMS

11.1. Explain why a smoke ring moves through the air.

11.2. Do the birds flying in Vee formation (other than the leader) have to supply asymmetric lift?   Explain.

11.3. Does the lead bird in a Vee formation need more or less power than he
would alone?

11.4. Show with a model how sweepback increases the dihedral effect.

REFERENCES

11.1. Abe Silverstein, S. Katzoff, and W. Kenneth Bullivant, Downwash and
    Wake behind Plain and Flapped Airfoils, NACA *TR* 651, 1939.

11.2. "Handbook of Chemistry and Physics," p. 191, Chemical Rubber Publishing Co.

11.3. H. Glauert, Some Applications of the Vortex Theory of Aerofoils, *R & M*
    752, 1921.

11.4. H. Glauert, Aerofoil Theory, *R & M* 723, 1921.

11.5. B. O. Pierce, "Short Table of Integrals," p. 41, Formula 300, Ginn & Company, Boston.

11.6. H. Kaden, Aufwicklung einer unstabilen Unstetigkeitsfache, *Ingenieur-
    Archiv*, Vol. II, No. 2, May, 1931, S 140–168.

11.7. H. Muttray, Investigations on the Amount of Downwash behind Rectangular and Elliptic Wings, *TM* 787, 1936.

11.8. John R. Spreiter and Alvin H. Sacks, The Rolling Up of the Trailing Vortex
    Sheet and Its Effect on the Downwash behind Wings, *Inst. Aeronaut.
    Sciences Preprint* 250, 1950.

CHAPTER 12

SPANWISE LOAD DISTRIBUTION FOR ARBITRARY WINGS

The problem of determining the spanwise distribution and hence the spanwise lift coefficient distribution has already been discussed in Chap. 10 through the use of Eq. (9.31), which relates the local downwash to a known distribution of circulation, and in that chapter a few cases embracing simple planforms were solved for both the circulation and downwash distributions. The case of the complete solution of an entirely arbitrary wing is a good deal more complicated, and usually some symplifying assumptions must be made.

Solutions to the problem of determining the spanwise lift distribution for arbitrary wings have been advanced by Multhopp (Ref. 12.1), Weissinger (Refs. 12.2 and 12.8), Faulkner (Ref. 12.9), Mutterperl (Ref. 12.10), Schrenk (Ref. 12.3), and others. Weissinger's work is an extension of Multhopp's theory, while Schrenk's approximation is a simple semi-graphical process based on the assumption that there is a tendency for the spanwise lift distribution of any wing to approach the ideal elliptical type. Multhopp's and Schrenk's methods will be discussed at some length in the pages to follow, it being believed that an understanding of Multhopp's method will suffice for the use of Weissinger's process.

First of all we recall that from a purely aerodynamic viewpoint the span loading curve should be elliptical in shape (Sect. 10.3) and that we can obtain the elliptic curve through the use of an elliptic planform with no aerodynamic twist or by properly twisting various planforms. A close approximation to an elliptic curve may be obtained through the use of $2\frac{1}{2}$ to 1 taper with no twist.

However, it must be noted that other factors may be more important to a particular design than minimum induced drag. Among these we find cost of construction and weight, and, indeed, the airplane designed to operate under conditions where the induced drag is small need not be concerned with small changes in induced drag at all.

While the minimum induced drag *for a given span* occurs with elliptic loading, great changes in induced drag are realized by changes in the span, and larger spans—until the wing weight becomes excessive—are in general preferable. Exceptions are noted for special criteria such as

245

maneuverability in roll (which increases as span decreases) and near-sonic drag rise (which may be decreased by sweepback). The location of the stall is another criterion that often influences the wing design.

Thus we see that many factors affect the wing layout, and it is quite necessary that the aerodynamicist be able to investigate an entire wing, learning beforehand what he may expect from the completed wing.

12.1. Multhopp's Spanwise Load Distribution Theory. All of the spanwise load distribution theories with which the author is familiar start with the well-known Prandtl downwash equation [Eq. (9.31)]

$$w(y_1) = \frac{1}{4\pi} \int_{-b/2}^{b/2} \frac{\frac{\partial \Gamma}{\partial y} dy}{y_1 - y} \tag{12.1}$$

where $w(y_1)$ = downwash velocity at y_1, ft per sec

y_1 = selected span station at which the downwash velocity is desired

y = location of the vortices causing the downwash

$\frac{\partial \Gamma}{\partial y} dy$ = the strength of the vortex being shed at y

However, the solution of Eq. (12.1) becomes complicated when $\frac{\partial \Gamma}{\partial y}$ is unknown (it was assumed in Chaps. 9, 10, and 11) since the downwash at a given point (which helps to determine the local angle of attack and hence the local lift) is a function of all the shed vortices, and these in turn are influenced by the circulation at the point in question. The fundamental relation is still that of Eq. (12.1), but a new setup is required to cover the situation when y is a variable. Multhopp also assumes that the wing may be replaced by a line vortex and writes Eq. (12.1) as

$$w(y) = \frac{1}{4\pi} \int_{-b/2}^{b/2} \frac{\frac{\partial \Gamma}{\partial \eta}}{y - \eta} d\eta \tag{12.2}$$

where y = spanwise station where the downwash velocity is desired (a change from the above)

η = spanwise station where vortices are located

The induced angle at y is then

$$\alpha_i(y) = \frac{1}{4\pi V} \int_{-b/2}^{b/2} \frac{\Gamma'(\eta)}{y - \eta} d\eta \tag{12.3}$$

where $\Gamma'(\eta) = \frac{\partial \Gamma}{\partial \eta}$.

In order to make the analysis nondimensional, we now write

$$\bar{y} = \frac{y}{b/2}; \qquad \bar{\eta} = \frac{\eta}{b/2}, \qquad \text{and} \qquad \gamma(\eta) = \frac{\Gamma(\eta)}{bV}$$

Also

$$d\bar{\eta} = \frac{2}{b}\,d\eta, \qquad \Gamma'(\bar{\eta}) = \frac{b}{2}\,\Gamma'(\eta), \qquad \text{and} \qquad \gamma'(\bar{\eta}) = \frac{b}{2}\,\frac{\Gamma'(\bar{\eta})}{bV}$$

Hence

$$\alpha_i(\bar{y}) = \frac{1}{2\pi} \int_{-1}^{+1} \frac{\left[\dfrac{b}{2}\dfrac{\Gamma'(\eta)}{bV}\right]\dfrac{2}{b}}{\dfrac{y}{b/2} - \dfrac{\eta}{b/2}}\, d\eta$$

$$= \frac{1}{2\pi} \int_{-1}^{1} \frac{\gamma'(\bar{\eta})}{\bar{y} - \bar{\eta}}\, d\bar{\eta} \tag{12.4}$$

From Eq. (4.26) we have the circulation at point $\bar{y}$,

$$\Gamma(\bar{y}) = \frac{c(\bar{y})Vc_l(\bar{y})}{2} \tag{12.5}$$

and using

$$c_l(\bar{y}) = a_0(\bar{y})[\alpha(\bar{y}) - \alpha_i(\bar{y})] \tag{12.6}$$

where $a_0(\bar{y})$ = two–dimensional slope of the lift curve at $\bar{y}$
$\quad\alpha(\bar{y})$ = angle of attack at $\bar{y}$ measured from zero lift
$\quad\alpha_i(\bar{y})$ = induced angle at $\bar{y}$
$\quad c_l(\bar{y})$ = section lift coefficient at $\bar{y}$
$\quad c(\bar{y})$ = chord at $\bar{y}$

we have

$$\Gamma(\bar{y}) = c(\bar{y})\,\frac{a_0}{2}\,(\bar{y})[\alpha(\bar{y}) - \alpha_i(\bar{y})]V$$

Again making Γ nondimensional and leaving off the $(\bar{y})$ on the right-hand side, where little confusion could occur, we have

$$\gamma(\bar{y}) = \frac{c}{b}\,\frac{a_0}{2}\,(\alpha - \alpha_i)$$

$$= \frac{a_0}{2\lambda}\,(\alpha - \alpha_i) \tag{12.7}$$

where λ = "local aspect ratio" = b/c.
Thus

$$\gamma(\bar{y}) = \frac{a_0}{2\lambda}\,\alpha - \frac{a_0\alpha_i}{2\lambda}$$

or

$$\frac{a_0}{2\lambda}\,\alpha = \gamma(\bar{y}) + \frac{a_0}{2\lambda}\,\alpha_i$$

Inserting Eq. (12.4), we find that the local angle of attack

$$\alpha = \frac{1}{2\pi} \int_{-1}^{1} \frac{\gamma'(\bar{\eta})}{\bar{y} - \bar{\eta}} \, d\bar{\eta} + \frac{2\lambda}{a_0} \gamma(\bar{y})$$

The local lift coefficient is, by Eq. (12.5),

$$c_l = \frac{2\Gamma}{cV}$$

or in nondimensional form

$$c_l = 2\lambda\gamma \tag{12.8}$$

where all quantities are at $\bar{y}$.

Hence, if γ is known as a $f(\bar{y})$, the local lift coefficient becomes known through Eq. (12.8).

Writing the equation for the total wing lift, we have

$$L = \int_{-1}^{1} q c c_l \, (\bar{y}) \, d\bar{y}$$
$$= q \int_{-1}^{1} 2c\lambda\gamma(\bar{y}) \, d\bar{y}$$

and

$$C_L = \frac{1}{S} \int_{-1}^{1} 2c\lambda\gamma(\bar{y}) \, d\bar{y} \tag{12.9}$$

The aspect ratio A.R. $= b^2/S$, and the total wing lift coefficient will be the same whether we integrate $\gamma(\bar{\eta}) \, d\bar{\eta}$ or $\gamma(\bar{y}) \, d\bar{y}$; so finally

$$C_L = \text{A.R.} \int_{-1}^{1} \gamma(\bar{\eta}) \, d\bar{\eta} \tag{12.10}$$

We may note at this time that the solution for γ to be later set up requires the values of α at several span stations. If there is no geometrical twist, α is the same at all stations. Under these circumstances, assuming $\alpha = 1$ radian will result in Eq. (12.10) yielding the lift curve slope $\frac{dC_L}{d\alpha}$ per radian. The induced drag will be

$$D_i = \int_{-b/2}^{b/2} \frac{w}{V} \, dL$$

and

$$C_{D_i} = \text{A.R.} \int_{-1}^{1} \gamma(\bar{\eta})\alpha_i \, d\bar{\eta} \tag{12.11}$$

The values of α_i will be found as follows:

$$\alpha_{\text{effective}} \equiv \alpha_e = \alpha - \alpha_i \tag{12.12}$$

Eq. (12.7) may be written

$$\gamma(\bar{y}) = \frac{a_0 \alpha_e}{2\lambda} \tag{12.13}$$

Since the values of γ will be found for certain stations along the span, the corresponding values of α_e may be found by Eq. (12.13). After finding C_{D_i} from Eq. (12.11), it will be of interest to compare it with Eq. (9.48), which is the ideal minimum.

Now then let us turn to the method of finding the local nondimensional circulation γ. From Eq. (12.5) the local circulation is a function of the local chord (which of course depends on the particular wing being considered) and the local lift coefficient. The latter may be considered a function of the local lift curve slope (which again depends on the particular wing being considered) and the local effective angle. The local effective angle is a function of the geometric twist (again depending on the wing being studied) and the local induced angle α_i.

The local induced angle is [Eq. (12.4)] a function of both the vortex locations $\bar{\eta}$ and the locations of the particular points $\bar{y}$ which combine to give the distance from the vortex to the point [h in Eq. (9.29)]. Multhopp's approach at this point is to find the local circulation at several *preselected* points instead of a completely general solution. Thus we shall only need γ at, say, 7 points. In turn we shall find c_l at 7 points. The beauty of this arrangement is that, if we agree to use the seven span-wise stations employed by Multhopp (a 15-point solution is also presented in Ref. 12.1), we may use his solutions for the $\bar{y} - \bar{\eta}$ distances; that is, all wings will be treated at similar fractions of the span.

A simple method to determine a satisfactory spanwise distribution

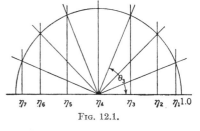

FIG. 12.1.

of our arbitrary points (we shall want more points near the tips, where the curve changes most rapidly) is shown in Fig. 12.1, where a semi-circle is emplaced upon the nondimensional wing and divided into eight equal angles θ according to the relation

$$\theta_\nu = \nu \frac{\pi}{m + 1} \qquad (12.14)$$

where m = number of stations selected—in our case 7

ν (nu) = identifying subscripts of a particular span station where the downwash is to be found—in our case any number from one to seven

The span station is then, from Fig. 12.1,

$$\bar{\eta}_\nu = \cos \theta_\nu \qquad (12.15)$$

Now then the local effective angle of attack at station ν, $\alpha_{e\nu}$, is equal to the local geometric angle of attack α_ν plus the effect of all the shed vortices $\Sigma\Delta\alpha_n$. Hence

$$\alpha_{e\nu} = \alpha_\nu + \sum_{n=1}^{7} \Delta\alpha_n \qquad (12.16)$$

where subscript ν signifies station under consideration and subscript n signifies vortex location.

For example, at station 1 we have

$$\alpha_{e1} = \alpha_1 + \Delta\alpha_1 + \Delta\alpha_2 + \Delta\alpha_3 + \Delta\alpha_4 + \Delta\alpha_5 + \Delta\alpha_6 + \Delta\alpha_7 \qquad (12.17)$$

The direct effect of the vortex at station 1 on the downwash at station 1 is, of course, zero, but it affects itself indirectly through its effect on the other vortices.

In order to get Eq. (12.16) into terms more easily handled with the wing geometry we note that from Eq. (12.13)

$$\frac{\alpha_e}{\gamma(\bar{y})} = \frac{2\lambda}{a_0}$$

or using the subscript notation, and defining a shorter form,

$$\frac{\alpha_e}{\gamma_\nu} \equiv b_\nu = \frac{2\lambda}{a_0}$$

The term $b_\nu\gamma_\nu$ is hence an angle, and we may write Eq. (12.16) as

$$\alpha_e = b_\nu\gamma_\nu = \alpha_\nu + \sum_{1}^{7} b_{\nu n}\gamma_n \qquad (12.18)$$

where $b_{\nu n}$ = a Multhopp coefficient tabulated in Table 12.1 accounting for the increment of induced angle at station ν due to vortex at station n, radians

b_ν = a Multhopp coefficient tabulated in Table 12.1 that accounts for the effect of the circulation at station ν on that at station $n = [2\lambda_\nu/a_0(\nu)] + b_{\nu\nu}$

In Multhopp's actual solution to obtain the values for the coefficients it develops that

$$b_{\nu\nu} = \frac{m+1}{4\sin\theta_\nu} \qquad (12.19)$$

$$b_{\nu n} = \frac{\sin\theta_\nu}{(\cos\theta_n - \cos\theta_\nu)^2}\frac{1-(-1)^{n-\nu}}{2(m+1)} \qquad (12.20)$$

from which we see that both $b_{\nu\nu}$ and $b_{\nu n}$ are always positive and $b_{\nu n} = 0$ when $n - \nu$ is an even number. We have at, say, station 2

Effective angle = geometric angle + induced angle

$$\alpha_{e2} = b_2\gamma_2 = \alpha_2 + \sum_1^7 b_{\nu n}\gamma_n \qquad (12.21)$$

from which

$$-b_2\gamma_2 + \sum_1^\infty b_{\nu n}\gamma_n = -\alpha_2$$

or, expanding,

$$-b_2\gamma_2 + b_{21}\gamma_1 + b_{22}\gamma_2 + b_{23}\gamma_3 + b_{24}\gamma_4 + b_{25}\gamma_5 + b_{26}\gamma_6$$
$$+ b_{27}\gamma_7 = -\alpha_2 \qquad (12.22)$$

Equation (12.22) may be written for all seven span stations, and if the stations are located as in Fig. 12.1, the distance, or "influence coefficients," $b_{\nu n}$ may be substituted from Multhopp's solution (Table 12.1).

TABLE 12.1. VALUES OF $b_{\nu\nu}$, $b_{\nu n}$ FOR $m = 7$

ν \ n	1	2	3	4	5	6	7
1	5.2262	1.8810	0	0.1464	0	0.0332	0
2	1.0180	2.8284	1.0972	0	0.0973	0	0.0180
3	0	0.8398	2.1648	0.8536	0	0.0744	0
4	0.0560	0	0.7887	2.000	0.7887	0	0.0560
5	0	0.0744	0	0.8536	2.1648	0.8398	0
6	0.0180	0	0.0973	0	1.0927	2.8284	1.0180
7	0	0.0332	0	0.1464	0	1.8810	5.2262

The seven equations may be reduced by noting that if the wing is symmetrical

$$\frac{2\lambda_1}{a_{01}} = \frac{2\lambda_7}{a_{07}} \qquad \alpha_1 = \alpha_7$$

$$\frac{2\lambda_2}{a_{02}} = \frac{2\lambda_6}{a_{06}} \qquad \alpha_2 = \alpha_6$$

$$\frac{2\lambda_3}{a_{03}} = \frac{2\lambda_5}{a_{05}} \qquad \alpha_3 = \alpha_5$$

so that

$$\gamma_1 = \gamma_7$$
$$\gamma_2 = \gamma_6$$
$$\gamma_3 = \gamma_5$$

The solution of the equations then reduces to that of four equations in four unknowns, the effect of the "missing" side being taken into account by combining appropriate coefficients. That is, we now have

$$b_{\nu n} + \sum_{1}^{\frac{7+1}{2}} B_{\nu n}\gamma_n = -\alpha_{\nu} \qquad (12.23)$$

where the coefficients $B_{\nu n}$ may be found in Table 12.2, and the complete set of equations is

$$\begin{aligned}
-b_1\gamma_1 + B_{12}\gamma_2 + B_{13}\gamma_3 + B_{14}\gamma_4 &= -\alpha_1 \\
B_{21}\gamma_1 - b_2\gamma_2 + B_{23}\gamma_3 + B_{24}\gamma_4 &= -\alpha_2 \\
B_{31}\gamma_1 + B_{32}\gamma_2 - b_3\gamma_3 + B_{34}\gamma_4 &= -\alpha_3 \\
B_{41}\gamma_1 + B_{42}\gamma_2 + B_{43}\gamma_3 - b_4\gamma_4 &= -\alpha_4
\end{aligned} \qquad (12.24)$$

TABLE 12.2. VALUES OF $B_{\nu n}$ FOR $m = 7$

ν \ n	1	2	3	4
1	0	1.9142	0	0.1464
2	1.0360	0	1.1944	0
3	0	.9142	0	0.8536
4	0.1121	0	1.5774	0

Example 12.1. Calculate the spanwise lift distribution for the following wing at $\alpha = 13.645°$:

Span $b = 200$ ft   No twist
Area $S = 4,000$ sq ft   Taper ratio = 3.0 to 1.0
Tip: Root:
  Chord $c_t = 10.0$ ft   Chord $c_r = 30.0$ ft
  Airfoil: NACA 0012   Airfoil: NACA 0018
$a_{0,\text{tip}} = 5.68$ per radian $a_{0,\text{root}} = 5.50$ per radian
Local chord $c_{\nu} = 30 - 20\bar{\eta}_{\nu}$

TABLE 12.3

1	2	3	4	5	6	7	8
Sta.	θ_{ν}	$\bar{\eta}_{\nu}$	$20\eta_{\nu}$	c_{ν}	a_0	2λ	$\dfrac{2\lambda}{a_0}$
1	22.50	0.9239	18.478	11.522	5.68	34.716	6.134
2	45.00	0.7071	14.142	15.858	5.65	25.224	4.486
3	67.50	0.3827	7.654	22.346	5.58	17.900	3.216
4	90.00	0.0000	0.000	30.000	5.50	13.333	2.424

Explanation of Table 12.3

1. Col. 1. ν varies from 1 to 4 for a 7-point symmetrical wing

2. Col. 2. $\nu \dfrac{\pi}{7+1}$ (57.3)

3. Col. 3. $\bar{\eta}_{\nu} = \cos\theta_{\nu}$

4. Col. 5.   Chord from formula for c_ν above
5. Col. 6.   Slope of the lift curve, interpolated between tip and root
6. Col. 7. $\lambda = \dfrac{b}{c}$

Substituting into Eq. (12.17), using Tables 12.1 and 12.2 and $\alpha_\nu = 1.0$ radian, we have

$$-(5.2262 + 6.1336)\gamma_1 + 1.9142\gamma_2 + 0.0 + 0.1464\gamma_4 = -1.0$$
$$1.036\gamma_1 - (2.8284 + 4.4866)\gamma_2 + 1.1944\gamma_3 + 0.0 = -1.0$$
$$0.00 + 0.9142\gamma_2 - (2.1648 + 3.2160)\gamma_3 + 0.8536\gamma_4 = -1.0$$
$$0.1121\gamma_1 + 0.00 + 1.5774\gamma_3 - (2.000 + 2.4242)\gamma_4 = -1.0$$

Solving

$$\gamma_1 = 0.1258$$
$$\gamma_2 = 0.1992$$
$$\gamma_3 = 0.2713$$
$$\gamma_4 = 0.32595$$

Finally, assuming that the spanwise lift coefficient distribution is desired for $\alpha = 13.645°$, we have the accompanying tabulation. Col. 5 is obtained from Eq. (12.8) after the values of γ found for 1.0 radian are multiplied by 13.645/57.3 to get them for 13.645°.

1	2	3	4	5
Sta.	$\bar{\eta}_\nu$	γ_ν	$2\lambda\gamma$	c_l
1	0.9239	0.1258	4.3673	1.040
2	0.7071	0.1992	5.0195	1.195
3	0.3827	0.2713	4.8563	1.156
4	0.0000	0.32595	4.3453	1.0347

Plots of γ and c_l are shown in Figs. 12.2 and 12.3.

The method of Multhopp is applicable to wings with twist and sweepback. Twist is simply introduced by varying the α terms in Eq. (12.17).

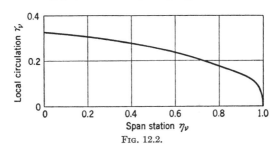

Fig. 12.2.

Sweepback is similar to twist and may be treated by a method given by Theilheimer (Ref. 12.6). However, it has been the experience of Flatt

254   *BASIC WING AND AIRFOIL THEORY*

(Ref. 12.4) and the author that the Schrenk method is so much shorter without any serious sacrifice of accuracy that sweepback examples or dis-

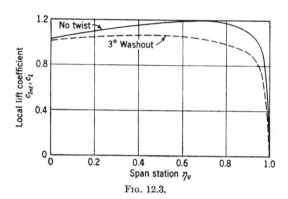

FIG. 12.3.

cussion of Multhopp or Weissinger will not be presented.   Twist is demonstrated in Ex. 12.2.

Example 12.2.   Find the spanwise load distribution for the wing of Ex. 12.1 if there is 3° aerodynamic twist (washout).   The simultaneous equations remain

1	2	3	4
Sta.	η_ν	α_{twist}, deg	α_{twist}, radians
1	0.9239	2.772	−0.04837
2	0.7071	2.121	−0.0370
3	0.3827	1.148	−0.0200
4	0.0000	0.000	0.000

the same as before on the left-hand side since the wing geometry is unaltered, but the twist is put into the geometric angle term on the right-hand side.

$$-(5.2262 + 6.1336)\gamma_1 + 1.942\gamma_2 + 0.00\gamma_3 + 0.1464\gamma_4 = +0.04837$$
$$1.036\gamma_1 - (2.8284 + 4.4866)\gamma_2 + 1.1944\gamma_3 + 0.0\gamma_4 = +0.037$$
$$0.0\gamma_1 + 0.914\gamma_2 - (2.1648 + 3.2160)\gamma_3 + 0.8536\gamma_4 = +0.020$$
$$0.1121\gamma_1 + 0.0\gamma_2 + 1.5774\gamma_3 + (2.000 - 2.4242)\gamma_4 = +0.0$$

Solving

$$\gamma_1 = -0.00544$$
$$\gamma_2 = -0.00664$$
$$\gamma_3 = -0.00516$$
$$\gamma_4 = -0.00198$$

If we now call the section lift coefficient for the untwisted case $c_{l_{nt}}$ and that for the twisted case c_{lt}, the total will be

$$c_l = c_{lnt} + c_{lt}$$

since the twist is referenced from the root-chord zero-lift line.
We have finally the accompanying tabulation.

1	2	3	4	5
Sta.	η_ν	$c_{lt} = 2\lambda\gamma$	c_{lnt}	c_l
1	0.9239	−0.189	1.040	0.851
2	0.7071	−0.1675	1.195	1.028
3	0.3827	−0.0924	1.156	1.064
4	0.00	−0.0264	1.0347	1.008

12.2. Schrenk's Method. Another approach to the problem of span-wise load distribution having much less theoretical foundation is presented by Flatt in Ref. 12.4. It follows a method first presented by Schrenk.

Schrenk's method makes allowance for the effect of the varying down-wash along the span of a nonelliptic wing by assuming that the final span load distribution for an untwisted wing is halfway between the actual planform shape and a semi-ellipse of the same area. As it turns out, the assumption is most reasonable. This section presents a method suitable for determining spanwise loading for simple arbitrary planforms and twists, but actually does not follow directly from theory.

Following the procedure used for examining chordwise pressures for thin airfoils, the spanwise lift distribution may be handled in two parts: the additional lift distribution cc_{la} and the basic lift distribution cc_{lb}. These distributions are considered separately and may be added to yield the net distribution. The additional part cc_{la} is that part of the lift due to angle of attack referenced from the wing zero lift angle. It is proportional to the average lift coefficient of the wing. When once determined for a wing lift coefficient $C_{L(1)}$, it may be found for $C_{L(2)}$ by

$$(cc_{la})_2 = \frac{C_{L2}}{C_{L1}} (cc_{la})_1 \tag{12.25}$$

The basic lift distribution is always constant for a given wing shape. It is the lift distribution when the wing $C_L = 0$ and exists only when the wing is twisted aerodynamically. The total lift

$$cc_l = cc_{la} + cc_{lb} \tag{12.26}$$

The theory assumes that the additional lift distribution cc_{l_a} for $C_L = 1.0$ is a mean between the actual distribution of wing chord and a semi-ellipse having the same area as the wing.

A second assumption is that the basic lift distribution is a mean between the wing zero lift line and the geometric-lift coefficient distribution due to twist.

Flaps and ailerons are treated as special cases of abrupt twist.

A third assumption (not covered in Ref. 12.4) is that the nondimensional spanwise lift distribution coefficient is decreased by a function of the sweepback angle so that the center of lift is moved outboard. This is demonstrated in Ex. 12.3.

The ideal distribution would be elliptic for the additional and zero for the basic.

It might be mentioned that the above assumptions fit in well with simple reasoning. First the flow is always apt to tend toward the path of least resistance (in this case elliptic loading), and second the sustentation of abrupt pressure variations in free air is quite foreign to practical experience.

The *additional lift* may be found as follows: Set the wing area S equal to one-half the area of an ellipse whose semi-minor axis is A and semi-major axis is $b/2$. Then

$$2S = \pi A \frac{b}{2}; \qquad A = \frac{4S}{\pi b} \tag{12.27}$$

The equation of the ellipse is then

$$\frac{x^2}{A^2} + \frac{y^2}{(b/2)^2} = 1.0$$

Letting y be out along the span and $x = c_e$ (the "elliptic" chord) be vertical, we have

$$\frac{(c_e)^2}{(4S/\pi b)^2} + \frac{y^2}{(b/2)^2} = 1.0$$

from which

$$c_l = \frac{4S}{\pi b}\sqrt{1 - \left(\frac{2y}{b}\right)^2} \tag{12.28}$$

From the assumption that $cc_{l_{a1}}$ is the mean between wing chord and the semi-ellipse having the same area as the wing, we have

$$cc_{l_{a1}} = \frac{c + c_e}{2} = \frac{c}{2} + \frac{2S}{\pi b}\sqrt{1 - \left(\frac{2y}{b}\right)^2}$$

where $cc_{l_{a1}}$ is the additional lift distribution when $C_L = 1.0$ (see Fig. 12.4).

For a variable thickness ratio use an effective chord c_1 defined by

$$c_1 = c \frac{a_0}{\bar{a}_0} \qquad (12.29)$$

where a_0 = local slope of the lift curve per degree
  $\bar{a}_0$ = average slope of the lift curve from

$$\bar{a}_0 = \frac{2}{S} \int_0^{b/2} c a_0 \, dy \qquad (12.30)$$

For straight taper wings both c and ca_0 vary linearly and

$$\bar{a}_0 = \frac{c_s a_{0s} + c_t a_{0t}}{c_s + c_t}$$

where the subscripts s and t refer to the plane of symmetry and tips, respectively.

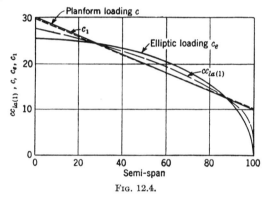

Fɪɢ. 12.4.

Correcting cc_{la1} for thickness, we have

$$cc_{la1} = \frac{c_1}{2} + \frac{2S}{\pi b} \sqrt{1 - \left(\frac{2y}{b}\right)^2} \qquad (12.31)$$

For $C_L = x$ instead of $C_L = 1.0$, we have

$$cc_{lax} = xcc_{la1}$$
$$= x\left[\frac{c}{2}\frac{a_0}{\bar{a}_0} + \frac{2S}{\pi b} \sqrt{1 - \left(\frac{2y}{b}\right)^2}\right]$$

The value of cc_{la} will be zero at the wing tip.

The *basic lift distribution* is found as follows:

Let ϵ_0 be the section twist angle referred to the zero lift line of the airfoil section at the plane of symmetry. The angle of zero lift for the wing is

$$\alpha_{ZL} = \int_0^{b/2} \epsilon_0 c \, dy \div \int_0^{b/2} c \, dy = \frac{2}{S} \int_0^{b/2} \epsilon_0 c \, dy \qquad (12.32)$$

The geometric twist angle ϵ_1 referred to the zero lift line of the wing is

$$\epsilon_1 = \epsilon_0 - \alpha_{ZL} \qquad (12.33)$$

Now then, the twist is not wholly effective. For instance, consider an untwisted wing and one twisted, say, 3° washout at the tip. The two distributions of circulation are shown in Fig. 12.3. Note that for equal total lifts the twisted wing must have higher circulation at the center-line, and hence the vortices shed before the tip region is reached have a greater total. Further, each of these shed vortices tends to *increase* the angle of attack at the wing tip. The amount by which they diminish the twist that made them is, of course, the question before us; and, it so happens, a value of 50 per cent is a very satisfactory approximation. Thus we could say that twist is half effective owing to the induced effect it creates.

Hence we may say

$$c_{lb} = \frac{\epsilon_1 a_0}{2} = \frac{\epsilon_0 - \alpha_{ZL}}{2} a_0$$

The net lift distribution for $C_L = x$ is therefore

$$cc_l = x \left[\frac{c}{2} \frac{a_0}{\bar{a}_0} + \frac{2S}{\pi b} \sqrt{1 - \left(\frac{2y}{b}\right)^2} \right] + \frac{(\epsilon_0 - \alpha_{ZL})}{2} a_0 c \qquad (12.34)$$

Two types of wing twist are considered:
1. Linear twist—ϵ_0 varies linearly along the semi-span
2. Uniform twist—$\epsilon_0 c$ varies linearly along the semi-span

An aerodynamic twist obtained by a change in airfoil section is usually assumed to be a linear twist. Uniform twist occurs whenever the tip chord is twisted relative to the root chord while retaining straight leading and trailing edges.

$$\epsilon_0 = \frac{2y}{b} \epsilon_t \qquad (12.35)$$

$$\epsilon_0 c = \frac{2y}{b} \epsilon_t c_t \qquad (12.36)$$

If the basic lift distribution is a function of more than one type of twist distribution, the net result can be obtained by the algebraic sum of the individual twist distributions for each joint.

For uniform twist along the entire span, the cc_{lb} and c_{lb} distributions are directly proportional to the twist angle. Thus if c_{lb1} is the distribution for ϵ_{t1},

$$c_{lbn} = nc_{lb} \text{ for } n\epsilon_t \qquad (12.37)$$

We may either plot values of c_l vs. span for the spanwise lift coefficient distribution; cc_l vs. span for the spanwise lift distribution; or $cc_l/\bar{c}C_L$ vs. span for the nondimensional spanwise lift distribution, where $\bar{c}$ is the mean geometric chord.

The effect of sweepback is simply found (see Ref. 12.7) from

$$\left(\frac{cc_l}{\bar{c}C_L}\right)_\Lambda = \left(\frac{cc_l}{\bar{c}C_L}\right)_{\Lambda=0} - \left(1 - \frac{2y}{b}\right)[2(1 - \cos \Lambda)] \qquad (12.38)$$

where Λ = angle of sweepback.

Additional items may be followed in Ex. 12.3, adapted from Ref. 12.4.

Example 12.3. Plot the spanwise distribution of lift coefficient and lift for the wing of Ex. 12.1 by the Schrenk method.

Span = 200 ft	Aspect ratio = 10	a_0 (root) = 0.096
Area = 4000 sq ft	Taper ratio = 3.0	a_0 (tip) = 0.099
Root airfoil: NACA 0018	Tip airfoil: NACA 0012	

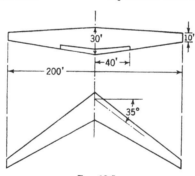

Fɪɢ. 12.5.

Aerodynamic twist: 3° washout from root to tip (straight leading and trailing edges, or uniform twist)

Consider (a) straight wing, flaps zero and (b) 35° sweepback, flaps zero.

From the airfoil dimensions an equation for the chord is set up, and then a table such as Table 12.4 may be filled in. For this particular type of problem careful slide-rule accuracy is probably sufficient.

From a plot of ca_0 vs. y, and Eq. (12.30), $\bar{a}_0 = 0.09675$.

The information from Table 12.4 is shown in Figs. 12.7 and 12.8. Possibly the most striking conclusions are:

1. The degree of dissimilarity between the c_l and cc_l distributions, caused, of course, by the distribution of chord. The load distribution is probably of more interest to the structural engineer, while the c_l distribution appeals largely to the aerodynamicist.

2. The effect of sweepback is to move the centroid of lift and the location of the start of the stall farther out.

12.3. Determination of the Stall Pattern. The information contained in Fig. 12.8 may be used to determine the approximate stall pattern by

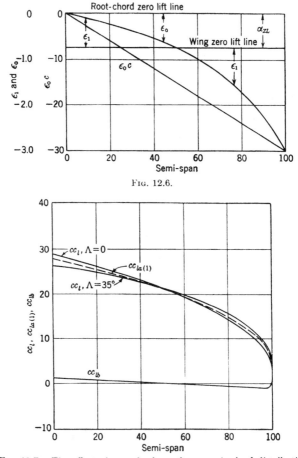

Fig. 12.6.

Fig. 12.7.   The effect of sweepback on the spanwise load distribution.

comparing the spanwise distribution of c_l with the spanwise distribution of $c_{l,\max}$. The $c_{l,\max}$ distribution will not be constant owing to the changing Reynolds number, and hence the stall may not start at the point where c_l is greatest.

The procedure embraces plotting the c_l and $c_{l,\max}$ distribution on one sheet and increasing wing C_L progressively until the c_l curve touches that of $c_{l,\,max}$. The spanwise location of the tangency is then the span-wise location of the start of the stall.

In general it is desirable to have the stall start somewhat out from the root but inboard of the innermost part of the aileron so that lateral con-

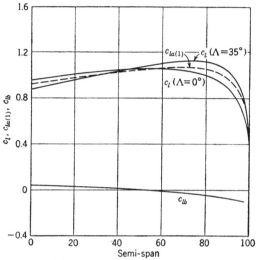

Fig. 12.8.   The effect of sweepback on the spanwise distribution of lift coefficient.

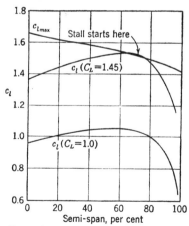

Fig. 12.9.   Determination of the spanwise location of the stall.

trol remains good.   A stall that is too strong at the root may cause undesirable tail buffeting.

The above method has not been very successful when applied to swept-back wings since it makes no provision for the three-dimensional effect

TABLE 12.4. SPANWISE LIFT DISTRIBUTION

1	2	3	4	5	6	7	8	9	10	11	12	13	14	15	16	17	18	19	20	21	22
Span sta. y, ft	c	c_l	$\sqrt{1-\left(\frac{2y}{b}\right)^2}$	c_e	$cc_{la(l)}$	$c_{la(l)}$	$\epsilon_0 c$	ϵ_0	ϵ_1	$\epsilon_1 c$	cc_{lb}	c_{lb}	cc_l	c_l	$\left(\frac{cc_l}{\partial C_L}\right)_{\Lambda=0}$	$1-\frac{2y}{b}$	F_Λ	$\left(\frac{cc_l}{\partial C_L}\right)_\Lambda$	$\left(\frac{cc_l}{\partial C_L}\right)_\Lambda$ $C_L=1.0$	c_l	cc_l
0	30.0	29.80	1.000	25.46	27.630	0.9210	0	0	0.750	22.50	1.090	0.0363	28.72	0.957	1.325	1.000	0.362	0.963	1.287	0.930	27.90
10.0	28.0	27.84	0.995	25.35	26.595	0.9498	−3.0	−0.107	0.643	18.00	0.872	0.0311	27.47	0.981	1.265	0.900	0.325	0.942	1.260	0.975	27.30
20.0	26.0	25.88	0.980	24.95	25.415	0.9775	−6.0	−0.231	0.519	13.50	0.654	0.0251	26.07	1.003	1.203	0.800	0.289	0.914	1.222	1.018	26.48
30.0	24.0	23.92	0.954	24.30	24.110	1.0050	−9.0	−0.375	0.436	9.00	0.436	0.0182	24.55	1.023	1.132	0.700	0.253	0.879	1.175	1.062	25.47
40.0	22.0	21.96	0.917	23.35	22.655	1.0298	−12.0	−0.545	0.205	4.50	0.218	0.0100	22.87	1.040	1.054	0.600	0.217	0.837	1.120	1.103	24.27
50.0	20.0	20.00	0.866	22.05	21.025	1.0513	−15.0	−0.750	0	0	0	0	21.03	1.051	0.969	0.500	0.181	0.788	1.054	1.142	22.82
60.0	18.0	18.04	0.800	20.37	19.205	1.0670	−18.0	−1.000	−0.250	−4.50	−0.218	−0.0121	18.99	1.055	0.874	0.400	0.145	0.729	0.973	1.170	21.08
70.0	16.0	16.08	0.714	18.20	17.140	1.0710	−21.0	−1.310	−0.560	−9.00	−0.436	−0.0272	16.70	1.044	0.769	0.300	0.108	0.661	0.882	1.193	19.10
80.0	14.0	14.12	0.600	15.29	14.705	1.0500	−24.0	−1.715	−0.965	−13.50	−0.654	−0.0460	14.05	1.004	0.647	0.200	0.072	0.575	0.768	1.188	16.64
90.0	12.0	12.16	0.430	11.10	11.630	0.9690	−27.0	−2.250	−1.500	−18.00	−0.872	−0.0726	10.76	0.897	0.495	0.100	0.036	0.459	0.612	1.105	13.25
95.0	11.0	11.18	0.312	7.95	9.565	0.8695	−28.5	−2.590	−1.840	−20.75	−0.982	−0.0890	8.58	0.781	0.395	0.050	0.018	0.377	0.503	0.991	10.90
97.5	10.5	10.69	0.222	5.65	8.170	0.7780	−29.3	−2.790	−2.040	−21.40	−1.037	−0.0985	7.13	0.680	0.329	0.025	0.009	0.320	0.427	0.882	9.25
100.0	10.0	10.20	0	0	0	0	−30.0	−3.000	−2.250	−22.50	0	−	−	−	0	0	0	0	0	0	0

Explanation of Table 12.4

1. Selected span stations
2. Local chord c from Fig. 12.4
3. Effective chord from Eqs. (12.29) and (12.30)
4. As shown
5. Elliptic chord from Eq. (12.28) $= 4S/\pi b$ (Col. 4)
6. cc_{la1} from Eq. (12.31)
7. Col. 6 ÷ Col. 2
8. $y/(b/2) \cdot (\epsilon_0 c)_{tip}$
9. Col. 8 ÷ Col. 2
10. α_{ZL} from Eq. (12.32) and Fig. 12.6; $\epsilon_1 = \epsilon_0 - \alpha ZL$
11. Col. 10 × Col. 2
12. From Eq. (12.34), using $\bar{a}_0 = 0.09675$ and Col. 11

13. Col. 12 ÷ Col. 2
14. Col. 6 + Col. 12
15. Col. 7 + Col. 13 or Col. 14 ÷ Col. 2
16. $C_L = 1.0$; $\bar{c} = MAC = \frac{2}{3}\left(c_{root} + c_{tip} - \frac{c_R c_T}{c_R + c_T}\right) = 21.67$ ft
17. As shown
18. F_Λ = Col. 17 × [2(1 − cos Λ)]
19. From Eq. (12.38)
20. Col. 19 was plotted and integrated, yielding $C_L = 0.749$
21. Col. 20 × $\bar{c}/c$
22. Col. 21 × Col. 2

of the boundary-layer flow. Currently no solution to this problem has been advanced, other than general experience to the effect that the stall usually appears early.

Example 12.4. Find the location of the start of the stall of a $\frac{1}{3}$ scale model of the wing used in Ex. 12.3. A stalling speed of 100 mph, standard air, is assumed.

1. The calculation of $c_{l,\,max}$ requires an additional reference, as the curves for variation of $c_{l,\,max}$ with Reynolds number are not included in this book. If Ref. 12.5 is not immediately available, one may consult nearly any of the 1946–1947 basic aerodynamic textbooks.

2. Table 12.2 and its column explanations are probably sufficient to explain the mechanism of working a simple stall problem. It should be noted, however, that c_l is not directly increased by increasing C_L but must be handled by Eqs. (12.2) and (12.10) since c_{lb} does not change with C_L.

TABLE 12.5

1	2	3	4	5	6	7	8	9
					$C_L = 1.0$		$C_L = 1.45$	
y	c	$\dfrac{RN}{10^6}$	$c_{l,\,max}$	c_{lb}				
					c_{lc}	c_l	c_{la}	c_l
0	10	9.35	1.66	0.036	0.921	0.957	1.335	1.371
10	9.34	8.73	1.635	0.031	0.950	0.981	1.376	1.407
20	8.66	8.10	1.616	0.025	0.978	1.003	1.416	1.441
30	8.00	7.48	1.600	0.018	1.005	1.023	1.456	1.474
40	7.34	6.86	1.585	0.010	1.030	1.040	1.483	1.493
50	6.66	6.22	1.565	0	1.051	1.051	1.524	1.524
60	6.00	5.61	1.546	−0.012	1.067	1.055	1.547	1.535
70	5.34	4.99	1.520	−0.027	1.071	1.044	1.553	1.526
80	4.67	4.37	1.490	−0.046	1.050	1.004	1.522	1.476
90	4.00	3.74	1.454	−0.073	0.969	0.897	1.405	1.322
95	3.67	3.43	1.435	−0.089	0.870	0.781	1.262	1.173
97	53.50	3.27	1.425	−0.099	0.778	0.680	1.127	1.028
100	3.33	3.12	—	—	—	—		

Explanation of Table 12.5

1. From Table 12.4, Col. 1
2. From Table 12.4, Col. 2
3. Reynolds number = 6,380 × 100 × 1.467 × c
4. $c_{l,\,max}$ from a cross plot of Fig. 28 of Ref. 12.5
5. From Table 12.4, Col. 13
6. From Table 12.4, Col. 7
7. Col. 5 + Col. 6
8. 1.45 × Col. 6
9. Col. 5 + Col. 8

REFERENCES

12.1. H. Multhopp, The Calculation of the Lift Distribution of Aerofoils, *RTP Translation* 3292, Durand Reprinting Committee.

12.2. J. Weissinger, The Lift Distribution of Swept-back Wings, *TM* 1120, March, 1947.

12.3. O. Schrenk, A Simple Approximation Method for Obtaining the Spanwise Lift Distribution, *TM* 948, 1940.

12.4. J. Flatt, Evaluation of Methods for Determining Spanwise Lift Distribution, *Army Air Force Tech. Rept.* 4952, 1943.

12.5. Eastman N. Jacobs and Albert Sherman, Airfoil Characteristics as Affected by Variations of the Reynolds Number, *TR* 586, 1937.

12.6. F. Theilheimer, Influence of Sweep on the Spanwise Lift Distribution of Wings, *JAS*, March, 1943.

12.7. Alan Pope and William R. Haney, Jr., Spanwise Lift Distribution for Sweptback Wings, *JAS*, August, 1949.

12.8. Nicholas H. Van Dorn and John De Young, A Comparison of Three Theoretical Methods of Calculating Span Load Distribution on Swept Wings, *TN* 1476, 1947.

12.9. V. M. Faulkner, The Calculation of Aerodynamic Loading on Surfaces of Any Shape, *R & M* 1910, 1943.

12.10. William Mutterperl, The Calculation of Span Load Distributions on Sweptback Wings, *TN* 834, 1941.

CHAPTER 13

MISCELLANEOUS APPLICATIONS OF PERFECT-FLUID THEORY

13.1. Types of Applications. The theory discussed previously has a multitude of other useful, practical applications. Among these we find biplane theory, effect of end-plate boundaries, ground effect, helicopter theory, and the corrections to indicated wind-tunnel data for spanwise loading effects, flow curvature, downwash, tail setting, hinge moments, rolling and yawing moments, and rotor and propeller data. It is beyond

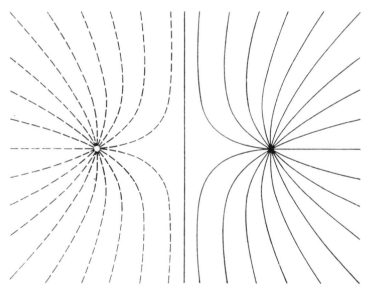

Fig. 13.1. Streamlines for a source near a wall.

the scope of this chapter to cover all these phenomena, but the basic principles of several will be discussed.

13.2. Single Boundaries. The case of a single boundary (the y axes, say) close to a source or a vortex is a simple starting place. For this instance we may emplace a mirror image of the source or vortex and, since it is below the y axis, assign its streamlines a negative value. Adding the two flows in a manner similar to that of Chap. 3 then yields a $\psi = 0$ boundary halfway between the real and image source. This is demonstrated for a source in Fig. 13.1. A similar procedure using vortices is followed for analyzing ground effect.

13.3. Solid Round Boundaries (Circular Wind-tunnel Jet). The case of a *small* wing in a closed circular jet offers an easy starting point for the study of wind-tunnel downwash corrections. Let us first consider the flow field produced in free air by the trailing vortices, as shown in Fig. 13.2. This flow, as indicated by the streamlines, cannot exist if we surround the wing by a solid tunnel boundary, and hence the normal

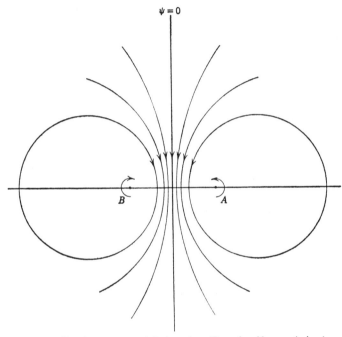

$\psi = 0$

Fig. 13.2. Streamlines for a vortex pair in free air. (*Reproduced by permission from " Wind Tunnel Testing" by Alan Pope, published by John Wiley & Sons, Inc., New York*, 1947.)

downward flow field must be altered by the tunnel walls. The problem is to find a series of sources, sinks, vortices, or doublets which, when combined with the free-air flow of Fig. 13.2, produce a $\psi = 0$ line which coincides with the actual tunnel wall. When this has been accomplished, it is a relatively simple matter to calculate the effect of the added items, which will be the same as the actual wall.

In the case of the small wing in the closed round jet a very simple solution arises, requiring only the addition of two vortices at a distance $x = \dfrac{R^2}{(b/2)}$ from the tunnel center, where R is the tunnel radius and $b/2$

the wing semi-span. The strength of these added vortices is the same as the original or bound vortices, but their direction is opposite (see Fig. 13.3).

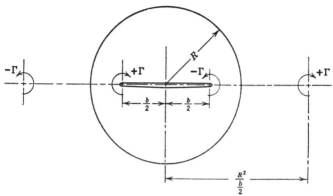

FIG. 13.3. The location for added vortices needed to simulate a round closed boundary about a wing. (*Reproduced by permission from "Wind Tunnel Testing" by Alan Pope, published by John Wiley & Sons, Inc., New York, 1947.*)

The addition of the two flows (Fig. 13.4) yields $\psi = 0$, exactly matching the tunnel wall.

Now consider the effect of the added vortices. Since they start at the lifting line of the wing, they are singly infinite and the velocity they induce (positive downward) at the wing centerline is [Eq. (9.28)]

$$w = -2\frac{\Gamma}{4\pi r} \qquad (13.1)$$

where r = distance from the vortex to the tunnel centerline.

Substituting $r = R^2/(b/2)$

$$w = -\frac{\Gamma}{2\pi R^2}\cdot\frac{b}{2} = -\frac{\Gamma b}{4\pi R^2} \qquad (13.2)$$

The strength of the added vortices equals that of the bound (wing) vortex and is

$$\Gamma = \frac{C_L S V}{2b}$$

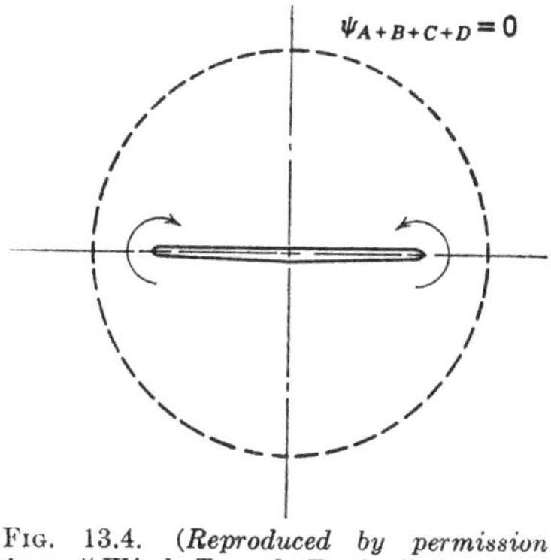

FIG. 13.4. (*Reproduced by permission from "Wind Tunnel Testing" by Alan Pope, published by John Wiley & Sons, Inc., New York, 1947.*)

Hence Eq. (13.2) becomes

$$w = -\frac{SVC_L}{4\pi R^2}\cdot\frac{b}{2b} = -\frac{1}{8}\frac{SV}{\pi R^2}C_L \qquad (13.3)$$

For small angles the downwash $\epsilon = w/V$, and writing the tunnel test section area $\pi R^2 = C$ we have

$$\Delta\alpha = \frac{w}{V} = -\frac{1}{8}\frac{S}{C}C_L \tag{13.4}$$

Since $\Delta\alpha$ is negative, the effect of the rigid boundaries is to decrease the free-air induced angle, making the tunnel angle of attack for a given C_L too small or making the wing appear to have a larger aspect ratio than it actually has. The *corrections* to the observed data will have the opposite sign from the above.

The actual coefficients become

$$\alpha = \alpha_{\text{tunnel}} + \delta\frac{S}{C}C_L \,(57.3) \tag{13.5}$$

and since

$$C_{D_i} = \frac{w}{V}C_L \tag{13.6}$$

$$C_D = C_{D,\text{tunnel}} + \delta\frac{S}{C}C_D{}^2 \tag{13.7}$$

where $\delta = 0.125$ for a small wing in a closed round jet. Other values of δ must be determined for other configurations. δ varies with model span, span load distribution, test section shape, and whether or not the wing is on the test section centerline.

It is a characteristic that closed test sections make the data appear optimistic, drag and angle for a given lift being smaller than in free air.

13.4. Open Round Boundaries. The basic closed-jet parameter of no lateral velocity through the tunnel wall (although lateral pressures may be sustained) is reversed for the case of the open jet: lateral velocities may then occur, but no lateral pressures can exist. This condition is met by the boundary being $\phi = 0$.

Again the small wing in the round test section proves a simple case, and again the condition $\phi = 0$ is met by the simple addition of two vortices at $x = R^2/(b/2)$. This time, however, the direction of rotation of the added vortices is identical with that of the nearest bound vortex. The mathematics are the same as before with the exception that the added velocity w is down, making the results appear pessimistic—too much drag at a given C_L, and too large an angle of attack appearing necessary.

The actual coefficients become

$$\alpha = \alpha_{\text{tunnel}} + \delta\frac{S}{C}C_L \,(57.3) \tag{13.8}$$

$$C_D = C_{D,\text{tunnel}} + \delta\frac{S}{C}C_L{}^2 \tag{13.9}$$

but in this case δ is negative.

13.5. Flat Boundaries. The problem of the tunnel with flat boundaries is more difficult than that of the small wing in a circular tunnel and requires an infinite series of vortex pairs. To demonstrate the principle, boundaries for a source between two vertical walls are shown in Figs. 13.5 and 13.6. It is seen that the walls are straightened up by the addition of each added image, and in a similar but more complicated manner this same process takes place for the wing vortex pair and its images.

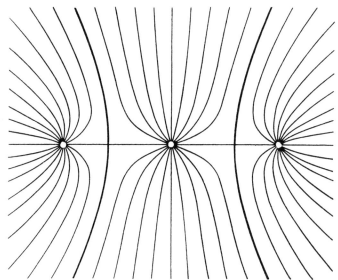

FIG. 13.5.   The streamlines for three equally spaced sources.

Using the same nomenclature as before, y along the span, but letting the tunnel breadth $= b_1$, we put an infinite series of vortex pairs along the y axis, spaced at $y = \pm m b_1$. Designate one of them by the letter A (Fig. 13.7). Owing to vortex pair A, which represents a wing at distance $m b_1$ from the tunnel centerline, the induced velocity at the tunnel centerline will be, according to Eq. (11.30),

$$\Delta w = - \frac{S V C_L}{8 \pi (m b_1)^2} \tag{13.10}$$

which will be upward.   The induced angle, positive up, will be*

$$\Delta(\Delta \alpha_1) = - \frac{\Delta w}{V} = \frac{S C_L}{8 \pi (m b_1)^2} \tag{13.11}$$

This amount is due to one vortex only.

---

* A positive w means that the induced angle is increased.   The additive correction $\Delta \alpha$ must then carry a negative sign.

FIG. 13.6.   The streamlines for five equally spaced sources.

Now while we must consider all the vortex pairs from $-\infty$ to $+\infty$ the one at 0, 0 is real, and its effect on itself is not part of the desirable

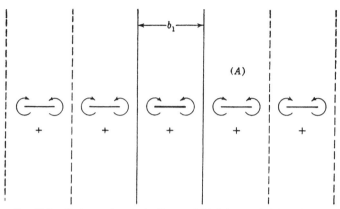

FIG. 13.7.   Vortex system to duplicate solid infinite vertical boundaries.

data.   Hence, neglecting the case of $m = 0$ as far as images go, we may write

$$\Delta\alpha_1 = \frac{SC_L}{8\pi b_1{}^2}\left(\sum_{-\infty}^{-1}\frac{1}{m^2} + \sum_{1}^{\infty}\frac{1}{m^2}\right) = \frac{2SC_L}{8\pi b^2}\sum_{1}^{\infty}\frac{1}{m^2} \qquad (13.13)$$

Now

$$\sum_{1}^{\infty}\frac{1}{m^2} = \frac{1}{1^2} + \frac{1}{2^2} + \frac{1}{3^2} + \cdots = \frac{\pi^2}{b} \qquad (13.14)$$

And hence

$$\Delta\alpha_1 = \frac{\pi S}{24b_1{}^2}\,C_L \qquad (13.15)$$

The case of the boundary above and below the wing follows from Eq. (11.31) and considerations of the flow along the z axis at large distances from the added "wings."   This time the spacing is $\pm nh$, but the image wings have alternately plus and minus signs being plus (like the bound wing) when n is even and minus when n is odd.   For the nearest image ($n = 1$) we have from Fig. 13.8

$$\Delta w = \frac{SVC_L}{8\pi z^2} = \frac{SVC_L(-1)^n}{8\pi(nh)^2} \qquad (13.16)$$

where $(-1)^n$ is inserted to satisfy the above sign requirements.

The total induced flow due to the images is

$$\Delta\alpha_2 = -\frac{w}{V} = -\frac{SC_L}{4\pi h^2}\sum_1^{\infty}\frac{(-1)^n}{n^2} \tag{13.17}$$

The series

$$\frac{(-1)^n}{n^2} = -\frac{\pi^2}{12} \tag{13.18}$$

and hence

$$\Delta\alpha_2 = \frac{\pi}{48}\frac{S}{h^2}C_L \tag{13.19}$$

From Eqs. (13.15) and (13.19) we see that the lateral vertical boundaries exert twice the influence of the horizontal boundaries.

13.6. Rectangular Boundaries. The problem of the effect of the walls of a rectangular wind tunnel on a small airfoil has been discussed by Glauert (Ref. 13.1), who found that a doubly infinite series of vortices is

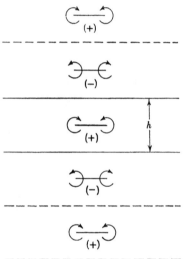

FIG. 13.8. Vortex system to duplicate solid infinite horizontal boundaries.

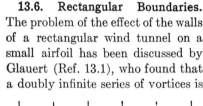
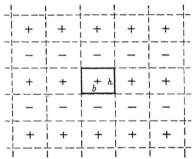

FIG. 13.9. Vortex field for simulating solid vertical and horizontal boundaries.

required and that they must be emplaced at the points $y = mb_1$ and $z = nh$, where m and n include all positive and negative integers. Considering only one image wing (at y, z) we find its effect on the bound wing to be [Eq. (11.31) and Fig. 13.9]

$$\Delta w = -\frac{1}{8\pi}\frac{y^2 - z^2}{(y^2 + z^2)^2}SVC_L \tag{13.20}$$

An image at (mb_1, nh) would yield the induced flow

$$\Delta w = -\frac{(-1)^n}{8\pi}\frac{m^2b_1^2 - n^2h^2}{(m^2b_1^2 + n^2h^2)^2}SVC_L \tag{13.21}$$

and for the entire field (m and $n = 0$ are considered later)

$$\Delta\alpha = -\frac{w}{V} = \frac{SC_L}{8\pi} \sum_{m=-\infty}^{\infty} \sum_{n=-\infty}^{\infty} (-1)^n \frac{m^2 b_1^2 - n^2 h^2}{(m^2 b_1^2 + n^2 h^2)^2} \quad (13.22)$$

Letting $\lambda = h/b_1$, we then have

$$\Delta\alpha = \frac{SC_L}{8\pi b_1^2} \sum_{m=-\infty}^{\infty} \sum_{n=-\infty}^{\infty} (-1)^n \frac{m^2 - \lambda^2 n^2}{(m^2 + \lambda^2 n^2)^2} \quad (13.23)$$

Unfortunately no simple value exists for the above expansion, and we must rearrange the summation as follows:

From trigonometry we may write

$$\cot z = \frac{1}{z} + 2z \sum_{1}^{\infty} \frac{1}{z^2 - m^2\pi^2} \quad (13.24)$$

Letting $z = i\lambda\pi x$ and recalling $\cot z = +i \coth iz$,

$$-i \coth \lambda\pi x = \frac{1}{i\lambda\pi x} - 2i\lambda\pi x \sum_{1}^{\infty} \frac{1}{m^2\pi^2 - (i\lambda\pi x)^2} \quad (13.25)$$

which clears to

$$\sum_{1}^{\infty} \frac{1}{m^2 + \lambda^2 x^2} = -\frac{1}{2\lambda^2 x^2} + \frac{\pi}{2\lambda x} \coth \lambda\pi x \quad (13.26)$$

Differentiating with respect to x,

$$\sum_{1}^{\infty} \frac{-2\lambda^2 x}{(m^2 + \lambda^2 x^2)^2} = \frac{1}{\lambda^2 x^3} + \frac{\pi}{2\lambda x}(-\operatorname{csch}^2 \lambda\pi x)\lambda\pi + \coth \lambda\pi x \left(\frac{-\pi}{2\lambda x^2}\right)$$

Now multiplying both sides by x,

$$\sum_{1}^{\infty} \frac{-2\lambda^2 x^2}{(m^2 + \lambda^2 x^2)^2} = \frac{1}{\lambda^2 x^2} - \frac{\pi^2}{2} \operatorname{csch}^2 \lambda\pi x - \frac{\pi}{2\lambda x} \coth \lambda\pi x \quad (13.27)$$

Adding Eqs. (13.26) and (13.27), we get

$$\sum_{1}^{\infty} \frac{(m^2 + \lambda^2 x^2) - 2\lambda^2 x^2}{(m^2 + \lambda^2 x^2)^2} = +\frac{1}{2\lambda^2 x^2} - \frac{\pi^2}{2} \operatorname{csch}^2 \lambda\pi x$$

Replacing x by n and multiplying both sides by $\sum_1^\infty (-1)^n$, we have for Eq. (13.23)

$$\sum_{m=1}^{\infty} \sum_{n=1}^{\infty} (-1)^n \frac{m^2 - \lambda^2 n^2}{(m^2 + \lambda^2 n^2)^2} = \frac{1}{2\lambda^2} \sum_{n=1}^{\infty} \frac{(-1)^n}{n^2} - \frac{\pi^2}{2}$$

$$\left[\sum_{n=1}^{\infty} (-1)^n \csc^2 \lambda \pi n \right] \quad (13.28)$$

Now we may write

$$\csc^2 x = \frac{4}{e^{2x} - 2 + e^{-2x}} \quad (13.29)$$

which after performing the indicated division becomes

$$\csc^2 x = 4 \left(\frac{1}{e^{2x}} + \frac{2}{e^{4x}} + \frac{3}{e^{6x}} + \cdots \right) = \sum_{p=1}^{\infty} 4p e^{-2px}$$

where p is any integer from 1 to ∞. Hence

$$\csc^2 \lambda \pi x = 4 \sum_{n=1}^{\infty} \sum_{p=1}^{\infty} p e^{-2\lambda \pi n p} \quad (13.30)$$

Substituting Eq. (13.30) into Eq. (13.28) and replacing the first term through the use of Eq. (13.18), we have

$$\sum_{m=1}^{\infty} \sum_{n=1}^{\infty} (-1)^n \frac{m^2 - \lambda^2 n^2}{(m^2 + \lambda^2 n^2)^2} = -\frac{\pi^2}{24} - 2\pi^2 \sum_{n=1}^{\infty} \sum_{p=1}^{\infty} (-1)^n p e^{-2\lambda \pi n p} \quad (13.31)$$

Equation (13.31) is summed for the range 1 to ∞ instead of $-\infty$ to ∞ as required by Eq. (13.23), and further manipulation is required. For a symmetric series we may write

$$\sum_{n=-\infty}^{\infty} A_n = 2 \sum_{n=1}^{\infty} A_n + A_{n=0} \quad (13.32)$$

and

$$\sum_{m=-\infty}^{\infty} B_m = 2 \sum_{m=1}^{\infty} B_m + B_{m=0} \quad (13.33)$$

The product of these two equalities yields

$$\sum_{m=-\infty}^{\infty} \sum_{n=-\infty}^{\infty} A_n B_m = 4 \sum_{m=1}^{\infty} \sum_{n=1}^{\infty} A_n B_m + A_{n=0} \cdot 2 \sum_{m=1}^{\infty} B_m$$

$$+ B_{m=0} \cdot 2 \sum_{n=1}^{\infty} A_n + A_{n=0} \cdot B_{m=0} \quad (13.34)$$

$$\sum_{m=-\infty}^{\infty} \sum_{n=-\infty}^{\infty} (-1)^n \frac{m^2 - \lambda^2 n^2}{(m^2 + \lambda^2 n^2)^2} = 4 \sum_{m=1}^{\infty} \sum_{n=1}^{\infty} (-1)^n \frac{m^2 - \lambda^2 n^2}{(m^2 + \lambda^2 n^2)^2}$$

$$+ 2 \sum_{1}^{\infty} \frac{m^2}{m^4} + 2 \sum_{1}^{\infty} (-1)^n \frac{-\lambda^2 n^2}{\lambda^4 n^4} \quad (13.35)$$

$$\sum_{m=-\infty}^{\infty} \sum_{n=-\infty}^{\infty} (-1)^n \frac{m^2 - \lambda^2 n^2}{(m^2 + \lambda^2 n^2)^2}$$

$$= \frac{\pi^2}{3} + 8\pi^2 \sum_{n=1}^{\infty} \sum_{p=1}^{\infty} (-1)(-1)^n p e^{-2\lambda \pi n p} \quad (13.36)$$

Expanding the n series of the term on the right-hand side, we have

$$\sum_{n=1}^{\infty} \sum_{p=1}^{\infty} (-1)(-1)^n p e^{-2\lambda \pi n p} = p \left(\frac{1}{e^{2\lambda \pi p}} - \frac{1}{e^{4\lambda \pi p}} + \cdots \right)$$

$$= p \frac{(e^{2\lambda \pi p} - 1)}{e^{4\lambda \pi p}}$$

$$= \frac{p}{e^{2\lambda \pi p} + 1}$$

Substituting back into Eq. (13.36), we have, finally,

$$\sum_{m=-\infty}^{\infty} \sum_{n=-\infty}^{\infty} (-1)^n \frac{m^2 - \lambda^2 n^2}{(m^2 + \lambda^2 n^2)^2} = \frac{\pi^2}{3} + 8\pi^2 \sum_{p=1}^{\infty} \frac{p}{1 + e^{2\lambda \pi p}} \quad (13.37)$$

Rewriting Eq. (13.23) in the form

$$\Delta\alpha = \delta \frac{S}{C} C_L = \frac{1}{8\pi} \frac{h}{b_1} \left(\frac{\pi^2}{3} + 8\pi^2 \sum_{1}^{\infty} \frac{p}{1 + e^{2\lambda \pi p}} \right) \frac{S}{h b_1} C_L$$

or

$$\Delta\alpha = \delta \frac{S}{C} C_L = \frac{h}{b_1} \left(\frac{\pi}{24} + \sum_{1}^{\infty} \frac{p}{1 + e^{2\lambda \pi p}} \right) \quad (13.38)$$

Fortunately the above series converges so rapidly for ordinary wind-tunnel dimensions that sufficient accuracy is obtained by solving for the first term only.   Thus for a height-width ratio of 1.0 (square test section)

$$\delta = \frac{1}{8\pi}\left(\frac{\pi^2}{3} + \frac{1}{1 + e^{2\pi}}\right)$$
$$= 0.131 + 0.006 = 0.137$$

Additional values are shown in Fig. 13.10.   Other tunnel configurations and larger models utilize the same principles as the above.

13.7. Apparent Additional Mass.   Another application of the fundamentals outlined previously develops when the acceleration of a body in a fluid is considered.   For instance, compare the state of a fluid in which a body at rest is immersed and one that has a body moving.   The moving body must push aside the fluid as it moves, and hence some of the fluid has motion and kinetic energy. Comparison between the state with the body at rest and in motion discloses that a difference of total energy exists.   If this fluid kinetic energy is divided by $\frac{1}{2}V^2$, where V is the velocity of the body in the direction of its motion, the dividend will

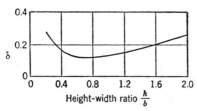

Fig. 13.10.   Values for δ for small wings in closed rectangular wind tunnels.

have the units of mass.   This "mass" is the *apparent additional mass* of the body in the direction of motion.   For the calculation of the force required to produce an acceleration of a body immersed in fluid the proper mass to use is that of the body plus its apparent mass or, as the sum of the two is defined, the *virtual* mass.   It now remains to consider the methods used to calculate the apparent additional mass.

The kinetic energy being stored in the fluid is a function of the velocity normal to the surface of the body, which by definition of ϕ may be written $\frac{\partial\phi}{\partial n}$.   Further, any irrotational motion has a function ϕ which may represent it, and in turn ϕ represents the impulse needed to create the flow. To understand this, recall that $\rho\phi$ has the units of *impulsive pressure*. The impulse may be found by multiplying $\rho\phi$ by the area of the body in question, S.

Now it is a proposition in dynamics that the work done by an impulse is equal to the product of the impulse and one-half the sum of the initial and final velocities.   If we let the initial velocity be zero, then the average velocity is $\frac{1}{2}\frac{\partial\phi}{\partial n}$, and the kinetic energy of the fluid due to a surface increment is

$$\Delta T = \rho\phi \cdot \frac{1}{2}\frac{\partial\phi}{\partial n}\,\Delta S$$

Then the total kinetic energy is

$$T = -\frac{\rho}{2}\int\int \phi\frac{\partial\phi}{\partial n}\,ds$$

The minus sign satisfies mathematical convention.
The apparent additional mass is then

$$m_A = \frac{2T}{V^2}$$

And the apparent additional volume is

$$K = \frac{m_A}{\rho}$$

Frequently the apparent additional volume is divided by the volume of the body to form a nondimensional coefficient sometimes called the *inertia factor* or the *coefficient of apparent additional volume:*

$$k = \frac{K}{\text{body volume}}$$

PROBLEMS

13.1. Plot the streamlines for a source ($Q = 200$) 10 ft from a wall.
13.2. Plot the streamlines for a vortex $\Gamma = 2\pi$ sq ft per sec 10 ft from a wall.
13.3. Plot the streamlines for a circular boundary, following the method of Sect. 13.3.
13.4. Using the Biot-Savart law (Chap. 9) and the work of Prandtl on pages 27 and 28 of *TR* 116, demonstrate that the weight of an airplane is represented by an increase of pressure on the ground.

REFERENCES

13.1. H. Glauert, The Interference of the Characteristics of an Airfoil in a Wind Tunnel of Rectangular Section, *R & M* 1459, 1932.
13.2. H. Glauert, Wind Tunnel Interference on Wings, Bodies and Airscrews, *R & M* 1566, 1933.
13.3. H. Glauert, "Aerofoil and Airscrew Theory," pp. 188–198, Cambridge University Press, London, 1926.

APPENDIX 1

The reason for selecting the transformation

$$z' = z\left(1 + i\Sigma\frac{A_na^n}{z^n}\right)$$

is not at all apparent, and in order to understand the thin-airfoil theory we shall have to consider Eq. (7.5) in detail.

In the first place the flow pattern due to two bodies having the same lift will be essentially the same at great distances from the bodies. Any transformation that relates the two should reduce to an equality when z is large. [Examine Eq. (5.3) for this property.] Thus the first thought for the z' transformation is a series of the form

$$z' = z + \frac{A_1}{z} + \frac{A_2}{z^2} + \frac{A_3}{z^3} + \cdots$$

but following through with this arrangement it is found that the proper end conditions are not simulated. We must, it develops, multiply the constants by i. In addition, since there is little difference between the circle and the near circle, the transformation will be close to 1.0. Hence we write

$$z' = z\left(1 + \frac{iA_1}{z} + \frac{iA_2}{z^2} + \frac{iA_3}{z^3} + \cdots\right)$$

The constants may be simplified and made smaller by writing them in terms of the radius of the circle as

$$z' = z\left(1 + \frac{iA_1a}{z} + \frac{iA_2a^2}{z^2} + \frac{iA_3a^3}{z^3} + \cdots\right) \tag{1}$$

because upon substitution of $z = ae^{i\phi}$ the a's cancel out and we get

$$z' = z(1 + iA_1e^{-i\phi} + iA_2e^{-2i\phi} + iA_3e^{-3i\phi} + \cdots)$$

Now ϵ is small so that $\cos\epsilon \cong 1$, and the product of $\sin\epsilon$ and small numbers may be neglected. Substituting $\phi = \theta + \epsilon$ and expanding, we get

$$z' = z\left(1 + \sum_1^\infty A_n\sin n\theta + i\sum_1^\infty A_n\cos n\theta\right) \tag{2}$$

Since

$$z' = a(1 + r)e^{i\theta}$$
$$= ae^{i\phi}(1 + r)e^{-i\epsilon}$$
$$= z(1 + r)e^{-i\epsilon}$$

we get, by expansion and remembering that r and ϵ are small,

$$z' = z(1 + r - i\epsilon)$$

and from Eq. (2)

$$r = \sum_1^\infty A_n \sin n\theta$$

$$\epsilon = -\sum_1^\infty A_n \cos n\theta$$

APPENDIX 2

Evaluation of the Integral of Eq. (8.28)

$$\epsilon_c = (\varphi - \theta)_c = -\frac{1}{2\pi} \int_0^{2\pi} \psi \cot \frac{\varphi - \varphi_c}{2} d\varphi \tag{1}$$

The integral of Eq. (8.28) may not be evaluated directly because, while $\psi = f(\varphi)$, no regularity exists that might make possible a mathematical substitution and hence a direct integration. We may not even plot the curve and measure the area beneath it with a planimeter since at $\varphi = \varphi_c$ the curve becomes discontinuous. However, that point is the only discontinuity, and if we set aside a region of width s on each side of φ_c, we may divide the remaining 350° (say) into strips and integrate each. The value of ψ for these strips may be taken as the arithmetic mean of the value at each strip edge. This mean, which we shall call ψ_A, is very nearly the average for the strip. Hence, for any strip from φ_1 to φ_2 not including φ_c, we have

$$\Delta\epsilon_c = -\frac{1}{\pi} \int_{\varphi_1}^{\varphi_2} \frac{1}{2} \psi_A \cot \frac{\varphi - \varphi_c}{2} d\varphi \tag{2}$$

$$= -\frac{1}{\pi} \psi_A \ln \sin \frac{\varphi - \varphi_c}{2} \Big]_{\varphi_1}^{\varphi_2}$$

When $\varphi_c > \varphi$, $(\varphi - \varphi_c)$ and its sine are negative and the meaning of the logarithm becomes obscure. However, this negative sign actually indicates only the relative position of φ as regards φ_c, and the proper relations are maintained if only the absolute value of the log term is retained. Thus,

$$\Delta\epsilon_c = -\frac{1}{\pi} \psi_A \ln \left| \frac{\sin \frac{\varphi_2 - \varphi_c}{2}}{\sin \frac{\varphi_1 - \varphi_c}{2}} \right| \tag{3}$$

The summation of all the $\Delta\epsilon_c$ terms will solve Eq. (1) for all but the narrow strip at φ_c which is considered now.

If the value of a function $y = f(x)$ is known at $x = a$, then the value of y at a point near a may be found from Taylor's theorem:

$$f(x) = f(a) + f'(a)\frac{(x-a)}{1!} + f''(a)\frac{(x-a)^2}{2!} \cdots f^r(a)\frac{(x-a)^n}{n!}$$

For our case we have to expand an $f(\varphi)$ in the region of φ_c. Hence, since

$$f(\varphi) = \psi \qquad\qquad f(\varphi_c) = \psi_c$$

$$f'(\varphi) = \frac{d\psi}{d\varphi} = \psi' \qquad f'(\varphi_c) = \psi_c'$$

$$f''(\varphi) = \frac{d^2\psi}{d\varphi^2} = \psi'' \qquad f''(\varphi_c) = \psi_c''$$

281

so that

$$\psi = \psi_c + (\varphi - \varphi_c)\psi_c' + \frac{(\varphi - \varphi_c)^2}{2}\,\varphi_c'' + \cdots \qquad (4)$$

Now Eq. (4) is valid only for a narrow strip (of width 2s) near φ_c, and under these conditions the integral of Eq. (3) becomes (omitting writing the $-1/2\pi$ term for a moment)

$$
\begin{aligned}
\epsilon_{\text{strip}} &= \int_{\varphi_c - s}^{\varphi_c + s} \psi \cot \frac{\varphi - \varphi_c}{2}\, d\varphi \\
&= \int_{\varphi_c - s}^{\varphi_c + s} \left[\psi_c + (\varphi - \varphi_c)\psi_c' + \frac{(\varphi - \varphi_c)^2}{2}\psi_c'' + \cdots \right] \cot \frac{\varphi - \varphi_c}{2}\, d\varphi \\
&= \int_{\varphi_c - s}^{\varphi_c + s} \psi_c \cot \frac{\varphi - \varphi_c}{2}\, d\varphi + \int_{\varphi_c - s}^{\varphi_c - s} \psi_c'(\varphi - \varphi_c) \cot \frac{\varphi - \varphi_c}{2}\, d\varphi \\
&\quad + \int_{\varphi_c - s}^{\varphi_c + s} \psi_c'' \frac{(\varphi - \varphi_c)^2}{2} \cot \frac{(\varphi - \varphi_c)}{2}\, d\varphi \quad (5)
\end{aligned}
$$

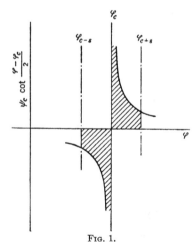

FIG. 1.

The advantage of Eq. (5) is that the values marked c are constants, and the integral is now in terms of one variable φ. It will develop that the second term is the only one that has any real value.

Consider the first term

$$\epsilon_{\text{strip}(1)} = \int_{\varphi_c - s}^{\varphi_c + s} \psi_c \cot \frac{\varphi - \varphi_c}{2}\, d\varphi \qquad (6)$$

It represents (see Fig. 1) a very small strip going to infinity along the positive axis when s is plus and oppositely when approached from the negative side. These opposite strips will therefore cancel out, and we see that the value of Eq. (6) for a small strip is zero.

The third term of Eq. (5)

$$\epsilon_{\text{strip}(3)} = \int_{\varphi_c - s}^{\varphi_c + s} \psi_c'' \frac{(\varphi - \varphi_c)^2}{2} \cot \frac{\varphi - \varphi_c}{2}\, d\varphi \qquad (7)$$

is in a way similar to Eq. (6), being plus when φ is slightly greater than φ_c and minus when φ is slightly less than φ_c. This integral does not go to infinity when $\varphi = \varphi_c$, however, but rather to the indeterminate $0 \cdot \infty$. However, the fact that the strips on each side of φ_c have opposite signs is all we need to conclude that the integral of Eq. (6) is also zero.

The second term

$$\epsilon_{\text{strip}(2)} = \int_{\varphi_c - s}^{\varphi_c + s} \psi_c' \cot \frac{\varphi - \varphi_c}{2} (\varphi - \varphi_c)\, d\varphi \qquad (8)$$

does not have the same plus and minus strips shown by the others. When $\varphi > \varphi_c$, we have a plus strip, but when $\varphi < \varphi_c$, both the cotangent and $\varphi - \varphi_c$ are negative, making the term positive. To simplify the integral, let

$$(\varphi - \varphi_c)/2 = u$$

Then $d\varphi = 2du$, and the limits of integration become $+s/2$ and $-s/2$. Our second term

$$\epsilon_{\text{strip}(2)} = 4\psi'_c \int_{-s/2}^{+s/2} u \cot u \, du \tag{9}$$

While Eq. (9) appears soluble by integration by parts, $\int u \cot u \, du$ is the "exception that proves the rule" and requires special treatment. In our case, however, an advantage accrues from having the integral strips very narrow, and so we may treat the area as a rectangle. First we must find the value of $u \cot u$ as $u \to 0$.

The rule requires that we rearrange the function until installation of the limit yields $\frac{0}{0}$. Hence

$$\lim_{u \to 0} \frac{u}{\tan u} = \frac{0}{0}$$

and replacing numerator and denominator by their derivatives and inserting the limit yields

$$\frac{1}{\sec^2 u} = 1.0$$

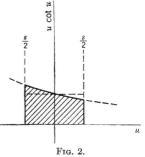

Fig. 2.

The shaded area in Fig. 2 becomes $1[(s/2) - (-s/2)] = s$, and the integral becomes

$$\epsilon_{\text{strip}} = 4\psi'_c s$$

And if our strip width be arbitrarily set at $-\pi/10$ to $+\pi/10$ ($s = \pi/10$),

$$\epsilon_{\text{strip}} = 4\psi_c' \frac{\pi}{10} = \frac{2\pi}{5} \psi_c'$$

Referring back to Eq. (8.28), the value of a strip is

$$\epsilon_{\text{strip}} = -\frac{1}{2\pi} \int_{\varphi_1}^{\varphi_2} \psi \cot \frac{\varphi - \varphi_c}{2} \, d\varphi$$

And this becomes (for the strip near φ_c)

$$\epsilon_{\text{strip}} = -\frac{1}{2\pi} \frac{2\pi}{5} \psi_c' = -\frac{\psi_c'}{5} \tag{10}$$

We are now in shape to integrate Eq. (8.28) for the complete range. We shall do this in strips of width $\pi/5$, using the average value of ψ, ψ_A, for a given strip width, and for that strip near φ_c we shall use the value from Eq. (10).

Now, from Eq. (3), the value of a strip (other than the one near φ_c) is

$$\Delta\epsilon_c = -\frac{1}{\pi}\psi_A \ln \left|\frac{\sin \dfrac{\varphi_2 - \varphi_c}{2}}{\sin \dfrac{\varphi_1 - \varphi_c}{2}}\right|$$

For the interval of ψ_A we select $\pi/5$, letting φ_c fall at the mid-point of one strip. We then have from Fig. 3 strips remaining as follows: $\varphi_c + (\pi/10)$ to $\varphi_c + (3\pi/10)$; $\varphi_c + (3\pi/10)$ to $\varphi_c + (5\pi/10)$; $\varphi_c + (5\pi/10)$ to $\varphi_c + (7\pi/10)$; $\varphi_c + (7\pi/10)$ to $\varphi_c + (9\pi/10)$; $\varphi_c + (9\pi/10)$ to $\varphi_c - (9\pi/10)$; $\varphi_c - (9\pi/10)$ to $\varphi_c - (7\pi/10)$; $\varphi_c - (7\pi/10)$ to $\varphi_c - (5\pi/10)$; $\varphi_c - (5\pi/10)$ to $\varphi_c - (3\pi/10)$; $\varphi_c - (3\pi/10)$ to $\varphi_c - (\pi/10)$.

The small strip for which the value of ϵ_c is $-\psi_c'/5$ is $\varphi_c - (\pi/10)$ to $\varphi_c + (\pi/10)$.

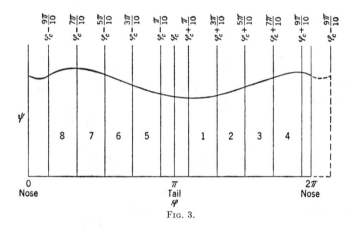

FIG. 3.

Let the value of ψ at $\varphi = \varphi_c + (\pi/10) = \psi_{\varphi_c+(\pi/10)}$. Then, writing out the summation series, we have from Eq. (3)

$$\epsilon_c + \frac{\psi_c'}{5} = -\frac{1}{\pi}\left[\frac{\psi_{\varphi_c+(\pi/10)} + \psi_{\varphi_c+(3\pi/10)}}{2} \ln \left|\frac{\sin (3\pi/20)}{\sin (\pi/20)}\right| \right.$$

$$+ \frac{\psi_{\varphi_c+(3\pi/10)} + \psi_{\varphi_c+(5\pi/10)}}{2} \ln \left|\frac{\sin (5\pi/20)}{\sin (3\pi/20)}\right| + \frac{\psi_{\varphi_c+(5\pi/10)} + \psi_{\varphi_c+(7\pi/10)}}{2} \ln \left|\frac{\sin (7\pi/20)}{\sin (5\pi/20)}\right|$$

$$+ \frac{\psi_{\varphi_c+(7\pi/10)} + \psi_{\varphi_c+(9\pi/10)}}{2} \ln \left|\frac{\sin (9\pi/20)}{\sin (7\pi/20)}\right| + \frac{\psi_{\varphi_c+(9\pi/10)} + \psi_{\varphi_c-(9\pi/10)}}{2} \ln \left|\frac{\sin (-9\pi/20)}{\sin (9\pi/20)}\right|$$

$$+ \frac{\psi_{\varphi_c-(9\pi/10)} + \psi_{\varphi_c-(7\pi/10)}}{2} \ln \left|\frac{\sin (-7\pi/20)}{\sin (-9\pi/20)}\right| + \frac{\psi_{\varphi_c-(7\pi/10)} + \psi_{\varphi_c-(5\pi/10)}}{2} \ln \left|\frac{\sin (-5\pi/20)}{\sin (-7\pi/20)}\right|$$

$$\left. + \frac{\psi_{\varphi_c-(5\pi/10)} + \psi_{\varphi_c-(3\pi/10)}}{2} \ln \left|\frac{\sin (-3\pi/20)}{\sin (-5\pi/20)}\right| + \frac{\psi_{\varphi_c-(3\pi/10)} + \psi_{\varphi_c-(\pi/10)}}{2} \ln \left|\frac{\sin (-\pi/20)}{\sin (-3\pi/20)}\right| \right]$$

Now the first and ninth terms above may be collected as

$$\frac{\psi_{\varphi c+(3\pi/10)} + \psi_{\varphi c+(\pi/10)}}{2} \ln \left| \frac{\sin (3\pi/20)}{\sin (\pi/20)} \right| + \frac{\psi_{\varphi c-(\pi/10)} + \psi_{\varphi c-(3\pi/10)}}{2} \ln \left| \frac{\sin (-\pi/20)}{\sin (-3\pi/20)} \right|$$

and since the sine of a minus angle may be written as minus the sine of the angle, the log terms above become identical, and we have as the average of the two coefficients

$$(\psi_{\varphi c+(\pi/5)} - \psi_{\varphi c-(\pi/5)}) \ln \left| \frac{\sin (3\pi/20)}{\sin (\pi/20)} \right| \tag{11}$$

In a similar manner the second and eighth, the third and seventh, the fourth and sixth may be combined. Writing

$$\psi_1 = \text{value of } \psi \text{ at } \varphi = \varphi_c + \frac{\pi}{5}$$

$$\psi_{-1} = \text{value of } \psi \text{ at } \varphi = \varphi_c - \frac{\pi}{5}$$

$$\psi_2 = \text{value of } \psi \text{ at } \varphi = \varphi_c + \frac{2\pi}{5}, \text{ etc.}$$

and completing the collections, we have

$$\epsilon_c = -\frac{1}{\pi} \left[\frac{\pi}{5} \psi_c' + (\psi_1 - \psi_{-1}) \ln \left| \frac{\sin (3\pi/20)}{\sin (\pi/20)} \right| + (\psi_2 - \psi_{-2}) \ln \left| \frac{\sin (5\pi/20)}{\sin (3\pi/20)} \right| \right.$$
$$\left. + (\psi_3 - \psi_{-3}) \ln \left| \frac{\sin (7\pi/20)}{\sin (5\pi/20)} \right| + (\psi_4 - \psi_{-4}) \ln \left| \frac{\sin (9\pi/20)}{\sin (7\pi/20)} \right| \right]$$

The interval $-9\pi/20$ to $+9\pi/20$ disappears as $\ln |-1| = 0$. Substituting values for the angles we have

$$\epsilon_c = -\frac{1}{\pi} [0.628\psi_c' + 1.0647(\psi_1 - \psi_{-1}) + 0.4431(\psi_2 - \psi_{-2}) + 0.2311(\psi_3 - \psi_{-3})$$
$$+ 0.1030 (\psi_4 - \psi_{-4})]$$

In a similar manner, a 20-point solution (Ref. 8.2) yields

$$\epsilon_c = -\frac{1}{\pi} [0.3142\psi_c' + 1.091(\psi_1 - \psi_{-1}) + 0.494(\psi_2 - \psi_{-2}) + 0.313(\psi_3 - \psi_{-3})$$
$$+ 0.217(\psi_4 - \psi_{-4}) + 0.158(\psi_5 - \psi_{-5}) + 0.115(\psi_6 - \psi_{-6}) + 0.0884(\psi_7 - \psi_{-7})$$
$$+ 0.0511(\psi_8 - \psi_{-8}) + 0.0251(\psi_9 - \psi_{-9})$$

but in this case

$$\psi_1 = \text{value of } \psi \text{ at } \varphi = \varphi_c + \frac{\pi}{10}, \text{ etc.}$$

Still further solutions are given in Refs. 8.4 and 8.5.

APPENDIX 3

Integration of

$$w = \frac{-\Gamma_0}{\pi b} \int_{-b/2}^{b/2} \frac{y \, dy}{(y_1 - y) \sqrt{b^2 - 4y^2}} \tag{1}$$

As previously noted, the denominator of the above expression vanishes at the points $y = y_1$, $b/2$, and $-b/2$, and we must therefore integrate the ranges $(-b/2) + \epsilon_1$ to $y_1 - \epsilon_2$, $y_1 + \epsilon_2$ to $(b/2) - \epsilon_3$ and finally obtain the complete value of the integral by taking the limits as ϵ_1, ϵ_2, and ϵ_3 approach zero.

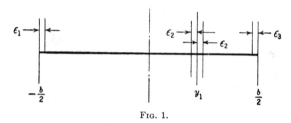

FIG. 1.

The problem may be simplified by making a trigonometric substitution,

$$y = -\frac{b}{2} \cos \theta$$

$$dy = \frac{b}{2} \sin \theta \, d\theta$$

$$y_1 = -\frac{b}{2} \cos \alpha$$

where α, which corresponds to the span station at which the downwash is sought, is held constant while the integration is being performed.

The new limits are 0 for $-b/2$ and π for $b/2$.

We now have

$$w = +\frac{\Gamma_0}{2\pi b} \int_0^\pi \frac{\cos \theta \, d\theta}{\cos \theta - \cos \alpha} \tag{2}$$

Dividing the numerator by the denominator, we get

$$w = \frac{\Gamma_0}{2\pi b} \int_0^\pi \left(1 + \frac{\cos \alpha}{\cos \theta - \cos \alpha} \right) d\theta \tag{3}$$

which becomes directly

$$w = \frac{\Gamma_0}{2b} - \frac{\Gamma_0 \cos \alpha}{2\pi b} \int_0^\pi \frac{d\theta}{\cos \theta - \cos \alpha} \tag{4}$$

286

Since the integral of the last term occurs under other circumstances, we shall take it out separately for consideration, writing

$$I = \int_0^\pi \frac{d\theta}{\cos\theta - \cos\alpha} \tag{5}$$

A further simplification occurs if we write $\tan(\theta/2) = x$. Then $\theta = 2\tan^{-1} x$ and $d\theta = 2/(1 + x^2)\,dx$. The integral becomes

$$I = \int_0^\infty \frac{\dfrac{2}{1 + x^2}\,dx}{\cos(2\tan^{-1}x) - \cos\alpha} = \frac{1}{\cos^2\dfrac{\alpha}{z}} \int_0^\infty \frac{dx}{\tan^2\dfrac{\alpha}{z} - x^2}$$

The integration of this discontinuous function [the integral becomes infinite when $\tan(\alpha/2) = x$] must be performed from 0 to $\tan(\alpha/2) - \epsilon$ and from $\tan(\alpha/2) + \epsilon$ to α, finally letting $\epsilon \to 0$. For proper evaluation it becomes expedient to change the signs of the second integral and to make the upper limit t, a large number. Then

$$I = \frac{1}{\cos^2\dfrac{\alpha}{2}}\left(\int_0^{\tan\frac{\alpha}{2} - \epsilon} \frac{dx}{\tan^2\dfrac{\alpha}{2} - x^2} - \int_{\tan\frac{\alpha}{2} + \epsilon}^t \frac{dx}{x^2 - \tan^2\dfrac{\alpha}{2}} \right) \tag{6}$$

Integrating, we have

$$I = \frac{1}{\cos^2\dfrac{\alpha}{2}}\left[\frac{1}{2\tan\dfrac{\alpha}{2}} \ln\frac{\tan\dfrac{\alpha}{2} + x}{\tan\dfrac{\alpha}{2} - x} \right]_0^{\tan\frac{\alpha}{2} - \epsilon} - \frac{1}{\cos^2\dfrac{\alpha}{2}}\left[\frac{1}{2\tan\dfrac{\alpha}{2}} \ln\frac{x + \tan\dfrac{\alpha}{2}}{x - \tan\dfrac{\alpha}{2}} \right]_{\tan\frac{\alpha}{2} + \epsilon}^t$$

which expands to

$$I = \frac{1}{2\sin\dfrac{\alpha}{2}\cos\dfrac{\alpha}{2}}\left(\ln\frac{2\tan\dfrac{\alpha}{2} - \epsilon}{\epsilon} + \ln 1 - \ln\frac{t - \tan\dfrac{\alpha}{2}}{t + \tan\dfrac{\alpha}{2}} + \ln\frac{\epsilon}{2\tan\dfrac{\alpha}{2} + \epsilon} \right)$$

Dropping the second term as equal to zero and combining the first and fourth, we have

$$I = \frac{1}{\sin\alpha}\left(\ln\frac{2\tan\dfrac{\alpha}{2} - \epsilon}{2\tan\dfrac{\alpha}{2} + \epsilon} - \ln\frac{1 - \dfrac{\tan\dfrac{\alpha}{2}}{t}}{1 + \dfrac{\tan\dfrac{\alpha}{2}}{t}} \right)$$

Passing to the limit as $\epsilon \to 0$ and $t \to \infty$, we see that the final value of the integral $I = 0$, or

$$I = \int_0^\pi \frac{d\theta}{\cos\theta - \cos\alpha} = 0$$

For the more general integral

$$I_n = \int_0^\pi \frac{\cos n\theta}{\cos\theta - \cos\alpha}\, d\theta$$

it is simplest to assume values for n and treat each term separately.   For instance, let $n = 2$.   We have then

$$I_2 = \int_0^\pi \frac{\cos 2\theta}{\cos\theta - \cos\alpha}\, d\theta = \int_0^\pi \frac{2\cos^2\theta - 1}{\cos\theta - \cos\alpha}\, d\theta$$

and by simple division

$$I_2 = \int_0^\pi 2\cos\theta\, d\theta + 2\cos\alpha \int_0^\pi \frac{\cos\theta}{\cos\theta - \cos\alpha}\, d\theta - \int_0^\pi \frac{1}{\cos\theta - \cos\alpha}\, d\theta$$

By direct integration the first integral drops out as equal to zero, and by Eq. (7) the last integral is zero, too.   The middle integral, by division, becomes

$$I_2 = 2\cos\alpha \int_0^\pi d\theta + 2\cos^2\alpha \int_0^\pi \frac{1}{\cos\theta - \cos\alpha}\, d\theta$$

of which the last integral is again zero.   Thus

$$I_2 = 2\pi\cos\alpha = \frac{2\pi\cos\alpha\sin\alpha}{\sin\alpha} = \frac{\pi\sin 2\alpha}{\sin\alpha}$$

By inspection, since we had selected $n = 2$, we have

$$I_n = \int_0^\pi \frac{\cos n\theta}{\cos\theta - \cos\alpha}\, d\theta = \frac{\pi\sin n\alpha}{\sin\alpha} \tag{8}$$

The student may easily check the validity of Eq. (8) by assuming values of n as above.

For the completion of Eq. (4), using Eq. (7), we see that

$$w = \frac{\Gamma_0}{2b} \tag{9}$$

ANSWERS TO PROBLEMS

1.1. $R = 3i + 3j - 5k$
$|R| = 6.56$

1.2. $R = 4i + 3j - 2k$
$|R| = 5.385$

1.3. $R = -4i + 4j + k$

1.4. $R = A \cdot B = -34$

1.5. M and N are parallel

1.6. $\theta = 170°$

1.7. $\alpha = 56.1°, \beta = 68.2°, \gamma = 138.0°$

1.8. $129.5°$

1.9. $-8i - 20j - 32k$

1.10. -24

1.11. $8i + 20j + 32k$

1.12. $iV_x \dfrac{\partial A_x}{\partial x} + iV_x \dfrac{\partial A_y}{\partial y} + iV_x \dfrac{\partial A_z}{\partial z}$ plus six more terms

1.14. $61.0°$

2.2. $\bar{z} = 3i - 2j$

3.1. Yes

3.2. Yes

3.3. $\psi = 50x - 86.5y$

3.4. $u = -4$ ft per sec, $q = 8.95$ ft per sec
$v = 8$ ft per sec, $\theta = 116°$

3.5. $u = 16$ ft per sec, $v = 21$ ft per sec
$q = 26.4$ ft per sec, $\theta = 52.7°$

3.6. $q_r = 7.14$ ft per sec, $q_\theta = -15.02$ ft per sec
$q = 16.62$ ft per sec, $\theta = 25.4°$

3.8. $q_r = 3.14$ ft per sec, $q_\theta = -2.47$ ft per sec
$q = 3.99$ ft per sec

3.9. 2.97 ft

3.10. $Q = 7600$ sq ft per sec

3.11. $q_{max} = 126$ ft per sec at $117°$ from x axis

3.12. $2,141, 2,117,$ and $2,039$ lb per sq ft

3.13. $\dfrac{\Delta p}{q} = 0.04$

3.14. 14.6 from horizontal

4.4. $\psi = 2xy$

4.5. $\phi = -2xy$

4.6. $q = 8.95$ ft per sec

4.7. Yes

4.8. Not irrotational

4.11. $-48.6°$

4.12. 137.5 ft per sec at $146°$ to horizontal

5.1. No

5.3. Usually not required

5.6. $c_l = 0.438, c_{m\frac{1}{4}} = 0$

5.7. $c_{m\frac{1}{4}} = -0.072$

5.8. $\alpha_{ZL} = -3.44°, c_{m\frac{1}{4}} = -0.0944$, C.P. $= -0.486$

5.9. 6.84 per radian, $1.095 \times 2\pi$

5.10. 0.894

5.11. NACA 4515

5.12. $\alpha_{ZL} = -4.59°, c_{m\frac{1}{4}} = -0.125$, C.P. $= -0.56$

9.1. 265 sq ft per sec

9.3. 5.09 ft per sec

9.4. 1.19 ft per sec

10.3. When wings are separate, the drag is doubled

10.4. (1) $2,660$; (2) $2,880$; (3) $2,730$

11.2. Yes

11.3. Less

INDEX

Chemistry

THE SCEPTICAL CHYMIST: THE CLASSIC 1661 TEXT, Robert Boyle. Boyle defines the term "element," asserting that all natural phenomena can be explained by the motion and organization of primary particles. 1911 ed. viii+232pp. 5⅜ x 8½.
0-486-42825-7

RADIOACTIVE SUBSTANCES, Marie Curie. Here is the celebrated scientist's doctoral thesis, the prelude to her receipt of the 1903 Nobel Prize. Curie discusses establishing atomic character of radioactivity found in compounds of uranium and thorium; extraction from pitchblende of polonium and radium; isolation of pure radium chloride; determination of atomic weight of radium; plus electric, photographic, luminous, heat, color effects of radioactivity. ii+94pp. 5⅜ x 8½.   0-486-42550-9

CHEMICAL MAGIC, Leonard A. Ford. Second Edition, Revised by E. Winston Grundmeier. Over 100 unusual stunts demonstrating cold fire, dust explosions, much more. Text explains scientific principles and stresses safety precautions. 128pp. 5⅜ x 8½.   0-486-67628-5

THE DEVELOPMENT OF MODERN CHEMISTRY, Aaron J. Ihde. Authoritative history of chemistry from ancient Greek theory to 20th-century innovation. Covers major chemists and their discoveries. 209 illustrations. 14 tables. Bibliographies. Indices. Appendices. 851pp. 5⅜ x 8½.   0-486-64235-6

CATALYSIS IN CHEMISTRY AND ENZYMOLOGY, William P. Jencks. Exceptionally clear coverage of mechanisms for catalysis, forces in aqueous solution, carbonyl- and acyl-group reactions, practical kinetics, more. 864pp. 5⅜ x 8½.
0-486-65460-5

ELEMENTS OF CHEMISTRY, Antoine Lavoisier. Monumental classic by founder of modern chemistry in remarkable reprint of rare 1790 Kerr translation. A must for every student of chemistry or the history of science. 539pp. 5⅜ x 8½.   0-486-64624-6

THE HISTORICAL BACKGROUND OF CHEMISTRY, Henry M. Leicester. Evolution of ideas, not individual biography. Concentrates on formulation of a coherent set of chemical laws. 260pp. 5⅜ x 8½.   0-486-61053-5

A SHORT HISTORY OF CHEMISTRY, J. R. Partington. Classic exposition explores origins of chemistry, alchemy, early medical chemistry, nature of atmosphere, theory of valency, laws and structure of atomic theory, much more. 428pp. 5⅜ x 8½. (Available in U.S. only.)   0-486-65977-1

GENERAL CHEMISTRY, Linus Pauling. Revised 3rd edition of classic first-year text by Nobel laureate. Atomic and molecular structure, quantum mechanics, statistical mechanics, thermodynamics correlated with descriptive chemistry. Problems. 992pp. 5⅜ x 8½.   0-486-65622-5

FROM ALCHEMY TO CHEMISTRY, John Read. Broad, humanistic treatment focuses on great figures of chemistry and ideas that revolutionized the science. 50 illustrations. 240pp. 5⅜ x 8½.   0-486-28690-8

Engineering

DE RE METALLICA, Georgius Agricola. The famous Hoover translation of greatest treatise on technological chemistry, engineering, geology, mining of early modern times (1556). All 289 original woodcuts. 638pp. 6¾ x 11. 0-486-60006-8

FUNDAMENTALS OF ASTRODYNAMICS, Roger Bate et al. Modern approach developed by U.S. Air Force Academy. Designed as a first course. Problems, exercises. Numerous illustrations. 455pp. 5⅜ x 8½. 0-486-60061-0

DYNAMICS OF FLUIDS IN POROUS MEDIA, Jacob Bear. For advanced students of ground water hydrology, soil mechanics and physics, drainage and irrigation engineering and more. 335 illustrations. Exercises, with answers. 784pp. 6⅛ x 9¼. 0-486-65675-6

THEORY OF VISCOELASTICITY (Second Edition), Richard M. Christensen. Complete consistent description of the linear theory of the viscoelastic behavior of materials. Problem-solving techniques discussed. 1982 edition. 29 figures. xiv+364pp. 6⅛ x 9¼. 0-486-42880-X

MECHANICS, J. P. Den Hartog. A classic introductory text or refresher. Hundreds of applications and design problems illuminate fundamentals of trusses, loaded beams and cables, etc. 334 answered problems. 462pp. 5⅜ x 8½. 0-486-60754-2

MECHANICAL VIBRATIONS, J. P. Den Hartog. Classic textbook offers lucid explanations and illustrative models, applying theories of vibrations to a variety of practical industrial engineering problems. Numerous figures. 233 problems, solutions. Appendix. Index. Preface. 436pp. 5⅜ x 8½. 0-486-64785-4

STRENGTH OF MATERIALS, J. P. Den Hartog. Full, clear treatment of basic material (tension, torsion, bending, etc.) plus advanced material on engineering methods, applications. 350 answered problems. 323pp. 5⅜ x 8½. 0-486-60755-0

A HISTORY OF MECHANICS, René Dugas. Monumental study of mechanical principles from antiquity to quantum mechanics. Contributions of ancient Greeks, Galileo, Leonardo, Kepler, Lagrange, many others. 671pp. 5⅜ x 8½. 0-486-65632-2

STABILITY THEORY AND ITS APPLICATIONS TO STRUCTURAL MECHANICS, Clive L. Dym. Self-contained text focuses on Koiter postbuckling analyses, with mathematical notions of stability of motion. Basing minimum energy principles for static stability upon dynamic concepts of stability of motion, it develops asymptotic buckling and postbuckling analyses from potential energy considerations, with applications to columns, plates, and arches. 1974 ed. 208pp. 5⅜ x 8½. 0-486-42541-X

METAL FATIGUE, N. E. Frost, K. J. Marsh, and L. P. Pook. Definitive, clearly written, and well-illustrated volume addresses all aspects of the subject, from the historical development of understanding metal fatigue to vital concepts of the cyclic stress that causes a crack to grow. Includes 7 appendixes. 544pp. 5⅜ x 8½. 0-486-40927-9

ROCKETS, Robert Goddard. Two of the most significant publications in the history of rocketry and jet propulsion: "A Method of Reaching Extreme Altitudes" (1919) and "Liquid Propellant Rocket Development" (1936). 128pp. 5⅜ x 8½.   0-486-42537-1

STATISTICAL MECHANICS: PRINCIPLES AND APPLICATIONS, Terrell L. Hill. Standard text covers fundamentals of statistical mechanics, applications to fluctuation theory, imperfect gases, distribution functions, more. 448pp. 5⅜ x 8½.
0-486-65390-0

ENGINEERING AND TECHNOLOGY 1650–1750: ILLUSTRATIONS AND TEXTS FROM ORIGINAL SOURCES, Martin Jensen. Highly readable text with more than 200 contemporary drawings and detailed engravings of engineering projects dealing with surveying, leveling, materials, hand tools, lifting equipment, transport and erection, piling, bailing, water supply, hydraulic engineering, and more. Among the specific projects outlined-transporting a 50-ton stone to the Louvre, erecting an obelisk, building timber locks, and dredging canals. 207pp. 8⅜ x 11¼.
0-486-42232-1

THE VARIATIONAL PRINCIPLES OF MECHANICS, Cornelius Lanczos. Graduate level coverage of calculus of variations, equations of motion, relativistic mechanics, more. First inexpensive paperbound edition of classic treatise. Index. Bibliography. 418pp. 5⅜ x 8½.   0-486-65067-7

PROTECTION OF ELECTRONIC CIRCUITS FROM OVERVOLTAGES, Ronald B. Standler. Five-part treatment presents practical rules and strategies for circuits designed to protect electronic systems from damage by transient overvoltages. 1989 ed. xxiv+434pp. 6⅛ x 9¼.   0-486-42552-5

ROTARY WING AERODYNAMICS, W. Z. Stepniewski. Clear, concise text covers aerodynamic phenomena of the rotor and offers guidelines for helicopter performance evaluation. Originally prepared for NASA. 537 figures. 640pp. 6⅛ x 9¼.
0-486-64647-5

INTRODUCTION TO SPACE DYNAMICS, William Tyrrell Thomson. Comprehensive, classic introduction to space-flight engineering for advanced undergraduate and graduate students. Includes vector algebra, kinematics, transformation of coordinates. Bibliography. Index. 352pp. 5⅜ x 8½.   0-486-65113-4

HISTORY OF STRENGTH OF MATERIALS, Stephen P. Timoshenko. Excellent historical survey of the strength of materials with many references to the theories of elasticity and structure. 245 figures. 452pp. 5⅜ x 8½.   0-486-61187-6

ANALYTICAL FRACTURE MECHANICS, David J. Unger. Self-contained text supplements standard fracture mechanics texts by focusing on analytical methods for determining crack-tip stress and strain fields. 336pp. 6⅛ x 9¼.   0-486-41737-9

STATISTICAL MECHANICS OF ELASTICITY, J. H. Weiner. Advanced, self-contained treatment illustrates general principles and elastic behavior of solids. Part 1, based on classical mechanics, studies thermoelastic behavior of crystalline and polymeric solids. Part 2, based on quantum mechanics, focuses on interatomic force laws, behavior of solids, and thermally activated processes. For students of physics and chemistry and for polymer physicists. 1983 ed. 96 figures. 496pp. 5⅜ x 8½.
0-486-42260-7

Mathematics

FUNCTIONAL ANALYSIS (Second Corrected Edition), George Bachman and Lawrence Narici. Excellent treatment of subject geared toward students with background in linear algebra, advanced calculus, physics and engineering. Text covers introduction to inner-product spaces, normed, metric spaces, and topological spaces; complete orthonormal sets, the Hahn-Banach Theorem and its consequences, and many other related subjects. 1966 ed. 544pp. 6⅛ x 9¼. 0-486-40251-7

ASYMPTOTIC EXPANSIONS OF INTEGRALS, Norman Bleistein & Richard A. Handelsman. Best introduction to important field with applications in a variety of scientific disciplines. New preface. Problems. Diagrams. Tables. Bibliography. Index. 448pp. 5⅜ x 8½. 0-486-65082-0

VECTOR AND TENSOR ANALYSIS WITH APPLICATIONS, A. I. Borisenko and I. E. Tarapov. Concise introduction. Worked-out problems, solutions, exercises. 257pp. 5⅜ x 8¼. 0-486-63833-2

AN INTRODUCTION TO ORDINARY DIFFERENTIAL EQUATIONS, Earl A. Coddington. A thorough and systematic first course in elementary differential equations for undergraduates in mathematics and science, with many exercises and problems (with answers). Index. 304pp. 5⅜ x 8½. 0-486-65942-9

FOURIER SERIES AND ORTHOGONAL FUNCTIONS, Harry F. Davis. An incisive text combining theory and practical example to introduce Fourier series, orthogonal functions and applications of the Fourier method to boundary-value problems. 570 exercises. Answers and notes. 416pp. 5⅜ x 8½. 0-486-65973-9

COMPUTABILITY AND UNSOLVABILITY, Martin Davis. Classic graduate-level introduction to theory of computability, usually referred to as theory of recurrent functions. New preface and appendix. 288pp. 5⅜ x 8½. 0-486-61471-9

ASYMPTOTIC METHODS IN ANALYSIS, N. G. de Bruijn. An inexpensive, comprehensive guide to asymptotic methods—the pioneering work that teaches by explaining worked examples in detail. Index. 224pp. 5⅜ x 8½ 0-486-64221-6

APPLIED COMPLEX VARIABLES, John W. Dettman. Step-by-step coverage of fundamentals of analytic function theory—plus lucid exposition of five important applications: Potential Theory; Ordinary Differential Equations; Fourier Transforms; Laplace Transforms; Asymptotic Expansions. 66 figures. Exercises at chapter ends. 512pp. 5⅜ x 8½. 0-486-64670-X

INTRODUCTION TO LINEAR ALGEBRA AND DIFFERENTIAL EQUATIONS, John W. Dettman. Excellent text covers complex numbers, determinants, orthonormal bases, Laplace transforms, much more. Exercises with solutions. Undergraduate level. 416pp. 5⅜ x 8½. 0-486-65191-6

RIEMANN'S ZETA FUNCTION, H. M. Edwards. Superb, high-level study of landmark 1859 publication entitled "On the Number of Primes Less Than a Given Magnitude" traces developments in mathematical theory that it inspired. xiv+315pp. 5⅜ x 8½. 0-486-41740-9

CATALOG OF DOVER BOOKS

CALCULUS OF VARIATIONS WITH APPLICATIONS, George M. Ewing. Applications-oriented introduction to variational theory develops insight and promotes understanding of specialized books, research papers. Suitable for advanced undergraduate/graduate students as primary, supplementary text. 352pp. 5⅜ x 8½.
0-486-64856-7

COMPLEX VARIABLES, Francis J. Flanigan. Unusual approach, delaying complex algebra till harmonic functions have been analyzed from real variable viewpoint. Includes problems with answers. 364pp. 5⅜ x 8½.
0-486-61388-7

AN INTRODUCTION TO THE CALCULUS OF VARIATIONS, Charles Fox. Graduate-level text covers variations of an integral, isoperimetrical problems, least action, special relativity, approximations, more. References. 279pp. 5⅜ x 8½.
0-486-65499-0

COUNTEREXAMPLES IN ANALYSIS, Bernard R. Gelbaum and John M. H. Olmsted. These counterexamples deal mostly with the part of analysis known as "real variables." The first half covers the real number system, and the second half encompasses higher dimensions. 1962 edition. xxiv+198pp. 5⅜ x 8½. 0-486-42875-3

CATASTROPHE THEORY FOR SCIENTISTS AND ENGINEERS, Robert Gilmore. Advanced-level treatment describes mathematics of theory grounded in the work of Poincaré, R. Thom, other mathematicians. Also important applications to problems in mathematics, physics, chemistry and engineering. 1981 edition. References. 28 tables. 397 black-and-white illustrations. xvii + 666pp. 6⅛ x 9¼.
0-486-67539-4

INTRODUCTION TO DIFFERENCE EQUATIONS, Samuel Goldberg. Exceptionally clear exposition of important discipline with applications to sociology, psychology, economics. Many illustrative examples; over 250 problems. 260pp. 5⅜ x 8½.
0-486-65084-7

NUMERICAL METHODS FOR SCIENTISTS AND ENGINEERS, Richard Hamming. Classic text stresses frequency approach in coverage of algorithms, polynomial approximation, Fourier approximation, exponential approximation, other topics. Revised and enlarged 2nd edition. 721pp. 5⅜ x 8½.
0-486-65241-6

INTRODUCTION TO NUMERICAL ANALYSIS (2nd Edition), F. B. Hildebrand. Classic, fundamental treatment covers computation, approximation, interpolation, numerical differentiation and integration, other topics. 150 new problems. 669pp. 5⅜ x 8½.
0-486-65363-3

THREE PEARLS OF NUMBER THEORY, A. Y. Khinchin. Three compelling puzzles require proof of a basic law governing the world of numbers. Challenges concern van der Waerden's theorem, the Landau-Schnirelmann hypothesis and Mann's theorem, and a solution to Waring's problem. Solutions included. 64pp. 5⅜ x 8½.
est mathematical minds–Pythagoras, Descartes, Euler, Pascal, Cantor, many more. Anecdotal, illuminating. 30 diagrams. Bibliography. 256pp. 5⅜ x 8½. 0-486-28973-7

A CONCISE HISTORY OF MATHEMATICS, Dirk J. Struik. The best brief history of mathematics. Stresses origins and covers every major figure from ancient Near East to 19th century. 41 illustrations. 195pp. 5⅜ x 8½.
0-486-60255-9

Physics

OPTICAL RESONANCE AND TWO-LEVEL ATOMS, L. Allen and J. H. Eberly. Clear, comprehensive introduction to basic principles behind all quantum optical resonance phenomena. 53 illustrations. Preface. Index. 256pp. 5⅜ x 8½. 0-486-65533-4

QUANTUM THEORY, David Bohm. This advanced undergraduate-level text presents the quantum theory in terms of qualitative and imaginative concepts, followed by specific applications worked out in mathematical detail. Preface. Index. 655pp. 5⅜ x 8½. 0-486-65969-0

ATOMIC PHYSICS (8th EDITION), Max Born. Nobel laureate's lucid treatment of kinetic theory of gases, elementary particles, nuclear atom, wave-corpuscles, atomic structure and spectral lines, much more. Over 40 appendices, bibliography. 495pp. 5⅜ x 8½. 0-486-65984-4

A SOPHISTICATE'S PRIMER OF RELATIVITY, P. W. Bridgman. Geared toward readers already acquainted with special relativity, this book transcends the view of theory as a working tool to answer natural questions: What is a frame of reference? What is a "law of nature"? What is the role of the "observer"? Extensive treatment, written in terms accessible to those without a scientific background. 1983 ed. xlviii+172pp. 5⅜ x 8½. 0-486-42549-5

AN INTRODUCTION TO HAMILTONIAN OPTICS, H. A. Buchdahl. Detailed account of the Hamiltonian treatment of aberration theory in geometrical optics. Many classes of optical systems defined in terms of the symmetries they possess. Problems with detailed solutions. 1970 edition. xv + 360pp. 5⅜ x 8½. 0-486-67597-1

PRIMER OF QUANTUM MECHANICS, Marvin Chester. Introductory text examines the classical quantum bead on a track: its state and representations; operator eigenvalues; harmonic oscillator and bound bead in a symmetric force field; and bead in a spherical shell. Other topics include spin, matrices, and the structure of quantum mechanics; the simplest atom; indistinguishable particles; and stationary-state perturbation theory. 1992 ed. xiv+314pp. 6⅛ x 9¼. 0-486-42878-8

LECTURES ON QUANTUM MECHANICS, Paul A. M. Dirac. Four concise, solutions in coverage of quantum mechanics, wave mechanics, angular momentum, molecular spectroscopy, more. 280 problems plus 139 supplementary exercises. 430pp. 6½ x 9¼. 0-486-65236-X

THEORETICAL SOLID STATE PHYSICS, Vol. 1: Perfect Lattices in Equilibrium; Vol. II: Non-Equilibrium and Disorder, William Jones and Norman H. March. Monumental reference work covers fundamental theory of equilibrium properties of perfect crystalline solids, non-equilibrium properties, defects and disordered systems. Appendices. Problems. Preface. Diagrams. Index. Bibliography. Total of 1,301pp. 5⅜ x 8½. Two volumes. Vol. I: 0-486-65015-4 Vol. II: 0-486-65016-2

WHAT IS RELATIVITY? L. D. Landau and G. B. Rumer. Written by a Nobel Prize physicist and his distinguished colleague, this compelling book explains the special theory of relativity to readers with no scientific background, using such familiar objects as trains, rulers, and clocks. 1960 ed. vi+72pp. 5⅜ x 8½. 0-486-42806-0

CATALOG OF DOVER BOOKS

A TREATISE ON ELECTRICITY AND MAGNETISM, James Clerk Maxwell. Important foundation work of modern physics. Brings to final form Maxwell's theory of electromagnetism and rigorously derives his general equations of field theory. 1,084pp. 5⅜ x 8½. Two-vol. set. Vol. I: 0-486-60636-8 Vol. II: 0-486-60637-6

QUANTUM MECHANICS: PRINCIPLES AND FORMALISM, Roy McWeeny. Graduate student-oriented volume develops subject as fundamental discipline, opening with review of origins of Schrödinger's equations and vector spaces. Focusing on main principles of quantum mechanics and their immediate consequences, it concludes with final generalizations covering alternative "languages" or representations. 1972 ed. 15 figures. xi+155pp. 5⅜ x 8½. 0-486-42829-X

INTRODUCTION TO QUANTUM MECHANICS With Applications to Chemistry, Linus Pauling & E. Bright Wilson, Jr. Classic undergraduate text by Nobel Prize winner applies quantum mechanics to chemical and physical problems. Numerous tables and figures enhance the text. Chapter bibliographies. Appendices. Index. 468pp. 5⅜ x 8½. 0-486-64871-0

METHODS OF THERMODYNAMICS, Howard Reiss. Outstanding text focuses on physical technique of thermodynamics, typical problem areas of understanding, and significance and use of thermodynamic potential. 1965 edition. 238pp. 5⅜ x 8½. 0-486-69445-3

THE ELECTROMAGNETIC FIELD, Albert Shadowitz. Comprehensive undergraduate text covers basics of electric and magnetic fields, builds up to electromagnetic theory. Also related topics, including relativity. Over 900 problems. 768pp. 5⅝ x 8¼. 0-486-65660-8

GREAT EXPERIMENTS IN PHYSICS: FIRSTHAND ACCOUNTS FROM GALILEO TO EINSTEIN, Morris H. Shamos (ed.). 25 crucial discoveries: Newton's laws of motion, Chadwick's study of the neutron, Hertz on electromagnetic waves, more. Original accounts clearly annotated. 370pp. 5⅜ x 8½. 0-486-25346-5

EINSTEIN'S LEGACY, Julian Schwinger. A Nobel Laureate relates fascinating story of Einstein and development of relativity theory in well-illustrated, nontechnical volume. Subjects include meaning of time, paradoxes of space travel, gravity and its effect on light, non-Euclidean geometry and curving of space-time, impact of radio astronomy and space-age discoveries, and more. 189 b/w illustrations. xiv+250pp. 8⅜ x 9¼. 0-486-41974-6

Paperbound unless otherwise indicated. Available at your book dealer, online at **www.doverpublications.com**, or by writing to Dept. GI, Dover Publications, Inc., 31 East 2nd Street, Mineola, NY 11501. For current price information or for free catalogues (please indicate field of interest), write to Dover Publications or log on to **www.doverpublications.com** and see every Dover book in print. Dover publishes more than 500 books each year on science, elementary and advanced mathematics, biology, music, art, literary history, social sciences, and other areas.